The Application of GIS Technologies in the Roman Period

Also available from Bloomsbury

Artifacts by Henry Hodges
Evidential Reasoning in Archaeology by Robert Chapman and Alison Wylie
GIS: Research Methods by Nick Bearman
Methods in Ancient Wine Archaeology: Scientific Approaches in Roman Contexts edited by Emlyn Dodd and Dimitri Van Limbergen

The Application of GIS Technologies in the Roman Period

Spatial Analysis, Historic Modelling and Reconstructions

Pedro Trapero Fernández

BLOOMSBURY ACADEMIC
LONDON • NEW YORK • OXFORD • NEW DELHI • SYDNEY

BLOOMSBURY ACADEMIC
Bloomsbury Publishing Plc
50 Bedford Square, London, WC1B 3DP, UK
1385 Broadway, New York, NY 10018, USA
29 Earlsfort Terrace, Dublin 2, Ireland

BLOOMSBURY, BLOOMSBURY ACADEMIC and the Diana logo are trademarks of Bloomsbury Publishing Plc

First published in Great Britain 2025

Cover image: An example of GIS analysis with visibility and mobility © Pedro Trapero Fernández

A catalogue record for this book is available from the British Library.

A catalog record for this book is available from the Library of Congress.

ISBN: HB: 978-1-3504-3370-0
ePDF: 978-1-3504-3371-7
eBook: 978-1-3504-3372-4

Typeset by RefineCatch Limited, Bungay, Suffolk

To find out more about our authors and books visit www.bloomsbury.com and sign up for our newsletters.

Contents

Illustrations

Tables

1

Introduction

Geographic Information Systems (GIS) are spatial tools that are specifically meant for land and environmental management, and have immense potential in various scientific disciplines. The use of these systems has been widespread in various scenarios, such as history and archaeology, where they play a crucial role in studying historical territories, identifying site locations, and addressing historical inquiries through the use of models. These models are representations of reality in which the main factors are analysed from theory, to be implemented in practice in one case study or another, with a high level of reproducibility.

Over the course of several decades, extensive research has been conducted on this subject, particularly concerning the practical application of GIS tools. It is essential to recognize that utilizing GIS for historical and archaeological purposes necessitates a comprehensive understanding of both the historical context and the technical aspects of the tool. Consequently, numerous books and manuals have been published throughout the years to provide guidance in this field.

One of the first relevant works, *Interpreting Space: GIS and Archaeology. Applications of Geographic Information Systems* (Allen et al. 1990), initiated discussions on a significant portion of these processes. From the 2000s onwards, there has been an exponential growth in studies on this topic, likely attributable to the widespread adoption of these tools among research institutions and the increased accessibility to spatial resources. Although we generate our own data, mainly from archaeological surveys, it also needs natural and topographical information. The spread of online resources and open access helps to create new and better models. Notable publications during this period include *Spatial Technologies in Archaeology* (Wheatley and Gillings 2002), *Landscape Archaeology and GIS* (Chapman 2006), *The Application of GIS in Landscape Archaeology* (Grau 2006), and *Geographical Information*

Systems in Archaeology (Conolly and Lake 2006). These transformations, in a way, from the analogical to the digital era, are succinctly captured in *Placing History: How Maps, Spatial Data, and GIS Are Changing Historical Scholarship* (Hillier and Knowles 2008).

Subsequently, numerous manuals have emerged, focusing on increasingly specific and significant issues. Examples of such works include *Computational Approaches to the Study of Movement in Archaeology…* (Polla and Verhagen 2014), *Best Practices of GeoInformatic Technologies for the Mapping of Archaeolandscapes* (Sarris 2015), and *Archaeology and Geomatics. Harvesting the Benefits of 10 Years of Training in the Iberian Peninsula* (Mayoral et al. 2016). The nature and specification of those works are increasingly the options to create models, normally based on specific case studies and the viability of open sources. In this matter, the development of those studies needs the scholarly tradition but also the accessibility of data.

In the context of the Hispanic region, we have an important improvement in recent years in parallel with the disposal of new and better information. For example, the LiDAR coverage of the whole territory favours studies, while the lack of this in other areas makes it difficult, as in the case of Portugal, being the same geographic entity. Compilation works of case studies have been published, such as the *Manual de tecnologías de la información geográfica aplicadas a la Arqueología* (Mínguez García and Capdevila 2016) and *GIS Applications in Roman Landscape and Territory: Methodologies and Models in Hispania* (Trapero and Carneiro 2023).

The evolution of GIS applications in history and archaeology continues in recent times, with methodological studies on new tools and emerging issues. Noteworthy publications in this regard include *Archaeological Spatial Analysis, A Methodological Guide* (Gillings et al. 2020), *Computational Approaches to Archaeological Spaces* (Bevan and Lake 2013), and *Archaeological 3D GIS* (Dell'Unto and Landeschi 2022).

However, despite the significant value these works hold in terms of methodology and serving as user manuals for GIS in this field, there are two major issues that need to be acknowledged. Firstly, they often consist of compilations of case studies conducted by different authors, where the methodology is briefly expressed, necessitating the consultation of external sources to fully comprehend the methods employed. Second, those works that are not compilations typically function as technical manuals on the usage of GIS, making it more challenging to extract historical content from them. With this understanding, the intention behind this work is to partially address this problem by offering a clear and concise methodology, centred on the Roman culture, which encompasses diverse spatial analysis scenarios susceptible to investigation, but with methods that are reproducible in other areas and chronologies.

The central issue in data analysis, particularly within the realm of GIS, lies in how the collected data is utilized and interpreted. In this regard, the primary focus of our work is precisely on this aspect, highlighting the uniqueness that our manuscript brings. Often, we encounter GIS analyses that present complex tools and impressive visualizations. However, a crucial question arises: What is the significance of all this? What is the underlying research question? What methodology was employed? How were the criteria measured and interpreted? While the creation of visually appealing maps through GIS is undoubtedly valuable, the potential of this technology extends far beyond mere cartography. Through analysis of information sources and their correlation with contemporary geography, GIS allows for the creation of models that accurately and intricately represent historical contexts. Through this involvement, we have observed that GIS is utilized in history and archaeology in numerous ways. This book is designed for specialists in GIS applied to history and archaeology, serving as a manual of methodologies and applications. However, it also caters to beginners interested in the advanced utilization of these tools, including students and other users involved in heritage-related fields.

These ranges go from simply creating thematic maps to utilizing them for geo-referenced database management and constructing spatial models that simulate historical realities. The latter aspect holds significant potential for advancing research in multiple areas, but it lacks a comparative and more unified methodology. In any scientific discipline, a model serves as a simplified representation of reality, enabling us to understand its fundamental elements and simulate various outcomes. Constructing a historical model necessitates more than just technical knowledge of GIS; it also requires an understanding of the mindset prevalent during the period under study. This is an important perspective that usually we do not show enough or compile in those studies, but a key question for investigation. This is why this book serves not just as a manual of GIS applications to history or archaeology. We want to show the method itself, using primarily our own case studies, but focusing the discussion on the problems of using ancient sources, the limitations of the tools, the possibilities for modifying the models and the need to create more examples to refine the precepts.

While we will explore the multifaceted uses of GIS in history and archaeology in detail in subsequent chapters, we will not delve extensively into its application within archaeology itself (Ejstrud 2003). This includes employing legacy data to reinterpret ancient excavated or surveyed sites (Casarotto 2022), leveraging the prospected materials, which have been the subject of numerous recent methodological publications (García Sánchez 2013; Grau 2016), or integrating data from new methodologies like geophysics applied to archaeology (Sarris et al. 2022). The nature of questions in those cases is more related to the understanding of cultural material than people's mindset. Of all the potential applications of these tools in recent times, it is the spatial-historical models that hold the greatest promise in current research, despite not necessarily being the most prevalent according to recent publications (Brughmans and Peeples 2017). The study of landscape and territory, along with its economic, social, and cultural implications, is not a new endeavour (Tukey 1977). Although these systems undoubtedly have certain limitations that render them somewhat questionable as models for certain researchers (Brouwer and Burg 2017), they have proven valuable in comprehending aspects such as settlement patterns (Bevan and Conolly 2006), interpreting extensive datasets (Cooper and Green 2016), understanding landscapes (Tilley 2010), and generating predictive models (Verhagen 2005).

This book primarily focuses on describing these methods and examining their use, particularly in the context of antiquity and Roman culture. The choice of Roman culture as the primary focus stems not only from the author's area of expertise but also from the fact that it represents a relatively homogeneous culture that spans diverse regions within the Mediterranean and Atlantic environments. Consequently, the applicability of the principles discussed extends to various countries. Additionally, Roman culture is one of the cultures for which we possess a wealth of technical information through scientific literature, such as the works of Roman agronomists (Martín 1971; Kolendo 1980) and surveyors (Dilke 1971; Chouquer 2007), as well as other relevant sources like legislation, which we explore in the corresponding chapter. Furthermore, Roman culture's temporal distance adds to the challenge of reconstructing its historical reality compared to other periods, such as the medieval or modern eras, where we have comparatively more quantitative and qualitative information at our disposal.

It is important to note that, while case studies are used as examples, the primary objective of this book is to summarize a set of methodological principles that can be applied universally, irrespective of location and potentially across different periods. Hence, we explore how to create historical models using GIS as a means to address historical questions. The methods presented in this book are not necessarily the best or the simplest, but they are predominantly based on two criteria (Scianna and Villa 2011). Firstly, they are complex models since they aim to represent not only simple inquiries but also historical realities that demand a comprehensive understanding of the period under study. Secondly, the methods employ similar approaches. In other words, by modifying the criteria and some of the tools, the process of reaching the theoretical approach remains largely similar, thereby enhancing reproducibility. Consequently, what is presented in this book are not rigid and flawless methodologies; rather, they represent open lines of research that, with slight parameter modifications, yield different results.

When conducting an analysis using GIS, we encounter several important considerations. First, we need to recognize that our analysis involves various geographic elements. In the context of landscape studies, our primary objective is to comprehend the changes in topography and economic factors over time. However, a critical methodological and conceptual challenge arises because we are not dealing with a singular, uniform landscape. In ancient times, we must acknowledge the existence of multiple landscapes, a fact substantiated by both archaeological findings and historical sources. Furthermore, urban landscapes are equally amenable to analysis using geographic information tools. It is important to emphasize that these models are not merely tools; they serve as a means to approach the reconstruction of historical realities. The applicability of each model is contingent upon the specific area we define in advance.

Defining these areas is not merely a matter of setting arbitrary boundaries but rather a comprehensive consideration of various factors. In this book, our primary focus lies on rural landscapes due to their inherent flexibility, enabling a more comprehensive examination. This perspective aligns with contemporary approaches to landscape and environmental studies, which offer pathways for reconstructing historical contexts.

Historiographically, the study of history is intertwined with the concept of space. The use of spatial models has a long historical tradition, particularly concerning the relationship between space, geography, and history. It encompasses a spatial component that can be examined from a diachronic perspective. This essentially involves the duality between space and time, where we analyse how various elements change over time or within a given space. Historical studies typically focus either on a specific moment in time or adopt a diachronic perspective.

In the aforementioned case, our primary objective is to gain a fundamental understanding of society during a particular historical period, such as the Roman era. However, it is essential to recognize the significant chronological variations within such historical periods. Our focus can range from a specific moment to subsequent periods. We focus on antiquity, but we express our models to be reproducible in other contexts such as prehistory, the Middle Ages, and the modern era. It indicates that the precepts and conditions used are referred to a Roman mentality, which is quite homogeneous and recognizable in different regions of what was the Roman Empire.

Our primary emphasis is on what we can consider as classical Roman culture. This encompasses the period from the end of the Roman Republic to the times of the Empire's crises. Our choice of chronology is not merely a matter of historical preference; it is also influenced by the cultural framework and prevailing mentalities. As we shall explore further, these factors shape the historical models we construct. Consequently, these models take into account changes in terrain conditions, as well as the specific circumstances of each province or case study.

The mentality of people during a specific historical moment can vary depending on their location within the Empire. However, our approach allows us to study the evolving behaviours of past societies over time. While our primary focus is on the Roman world, the principles and methodologies we employ can be applied to other contexts beyond our immediate scope.

Our perspective entails a duality of concepts. On one hand, it seeks to understand the mentality of a particular historical era from a societal viewpoint. On the other hand, it involves understanding the geographical determinants that can be identified within that mentality. These determinants are amenable to modelling, although it is important to note that not all the historical models we develop are universally applicable without modifying the prevailing mentality of the time.

For example, when calculating a suitable space for cultivation, we take into account the direction of detrimental winds, which in principle may remain similar over time, barring climatic changes that must be considered. However, there are also regional differences, as in the case of Roman *Baetica*, where Columela mentions a detrimental wind from the East, contrary to the orientation generally accepted in the rest of the Empire (*Rust.* 3.12.6). Understanding these peculiarities and determining appropriate criteria is what enables the general method discussed, with Roman examples, to be reproducible in other contexts and chronologies.

Our approach aims to bridge the gap between historical mindset and modern technology, primarily employing spatial modelling techniques. While these tools offer powerful analytical capabilities, our success as researchers ultimately depends on our ability to formulate meaningful historical questions and effectively apply these tools while respecting the historical context and mindset.

In this book, we compile a collection of the most successful types of spatial models, discussing their limitations and potential for implementation in other areas. Subsequently, we conduct a thorough critical analysis of each method and its underlying criteria, presenting them in an easy way to enable other researchers or students wanting to be specialists in the field, to replicate these approaches. To facilitate this, we provide our own specifically designed examples throughout the book. We augment the examples with a comprehensive graphical section, incorporating figures, graphs, and tables that exemplify and simplify the technical aspects of the process. Furthermore, we provide a critical review of the up-to-date bibliography. While the examples predominantly originate from our own research for the sake of clarity and the liberty to criticize our own approaches, we also draw upon a vast array of case studies from the existing literature, particularly those that address complex historical questions, which serve as excellent exemplars (Van Leusen 2002; Goodchild 2007; Viitanen 2010).

The structure of this book comprises three main introductory parts and five explanations of models. The first part focuses on methodological aspects, introducing historical models with GIS, and emphasizing the importance of mindset and good practices. This initial discussion is pertinent because we want to clarify who this book is intended for and the methodological need to start from historical questions before using GIS tools.

The second part delves into technical issues related to GIS tools. While the intention is not to develop a comprehensive technical GIS manual, it aims to provide descriptions of commonly used tools and offer tips for people who want to specialize in GIS in history as well as application tricks for more experienced users. This is necessary because of the small details and the method used to modify and construct spatial data. Usually, the hardest part of the model is the access and transformation of those precepts. Because of this, we will write about GIS principles, useful tools, advanced processes and other approaches to GIS studies.

The third part discusses the resources and methods of geographical origin that can be utilized, along with guidance on accessing them. Although those datasets would depend on the country and scale of work, we will describe the useful ones, how they usually appear in online resources, how to transform them to integrate with our model and finally how to create our own data with GIS tools.

Lastly, a series of chapters present specific applications of different models that can be employed. These chapters follow a consistent structure, defining the historical problem, detailing the factors to consider, explaining the methodology for modelling, and providing examples of application. The first one is about the land use models, with three variations depending on our expectations. Those are focused on the calculation of crop areas, economic uses and production excess to understand the economy of a certain rural space. The second group of models is about mobility. In this case, four variations depending on land, river, sea and increased economic cost of transport are explained, with fuller examples and difficulties of integrating the historical questions. As we did in the previous section, we concentrate on one particular tool, visibility, to tackle the inter-visibility and defensibility of sites. We continue the last two chapters focusing on topographical reconstruction and predictive models, both with two variations of the method. Finally, we devote a final chapter to discussing perspectives, how those models would evolve and the implications and lines for their development.

2

GIS Applied to History and Archaeology

Geographic Information Systems (GIS), as geo-referenced databases, are utilized for the management, analysis, and visualization of spatial data. Anything with a spatial dimension can be represented and studied through GIS. Consequently, these tools initially emerged from managing natural spaces or civil engineering, particularly in urban, transportation, or rural planning. Although their application to other domains is relatively recent, they offer similar potential. However, what are the implications when tools designed for contemporary applications retroactively apply to earlier historical periods? This chapter addresses this question, focusing on historical inquiries and distinguishing between various models.

2.1 Transition from Traditional Methods to Computerization

When discussing digital history or archaeology, many immediately associate them with 3D models and advanced graphics, viewing them as a means to showcase and augment the value of heritage. While this aspect is indeed significant, there exist numerous other applications and utilities. Technology has become an integral component of our daily lives, permeating scientific disciplines such as history and archaeology, which are not exempt from its influence.

It often occurs that these 'new' technologies engender feelings of obsolescence or quickly lose their novelty. However, Geographic Information Systems might be an exception, as they are widely utilized and accepted within the research community. Although they constituted a novel methodology a few decades ago, their prevalence has increased more recently. While perceived by many as primarily a geo-referenced database, a GIS should, for us, serve as a platform for posing inquiries, formulating methodologies, testing scenarios, and ultimately generating new insights. Throughout this process, it is crucial to establish a historical foundation, a consideration that is sometimes overlooked.

We have swiftly transitioned from relying on books and fieldwork to embracing spatial and digital studies, often with mixed success. It is not about assigning blame but rather acknowledging that in the current paradigm, we encounter a significant contradiction arising from specialized work (Haciguzeller 2012). The technical expertise necessary for GIS analytical processes means that few historians or archaeologists consider utilizing these tools, often leaving the task to specialists in the field. This is not to say that they do not use it; today, it is almost an obligatory tool, even if only for cartographic representation.

In general, we do not conflate the roles of butchers and shoemakers, nor do we expect one to instruct the other on their craft. However, it is both possible and beneficial to collaborate in a harmonious interdisciplinary approach. While this method is effective in many cases, it has led to a dearth of essential multidisciplinary studies. Ideally, the individual conducting GIS processes and the one comprehending the historical problem should either be the same or collaborate seamlessly.

Over time, this issue has gradually evolved to our current state. The democratization of GIS tools through the internet, particularly through free software, means that anyone with a computer can access and utilize them for their own processes. However, it is worth noting that these available tools were not specifically designed for historical or archaeological studies but rather for more generic applications, often related to land and resource management. What does this signify?

It suggests that we have vast potential at our disposal, yet the tools were not tailored precisely to our requirements. While it is feasible to use a 'knife' for various tasks related to our expertise, it cannot match the efficiency of a dedicated 'screwdriver'. The utilization of these non-specialized tools in our field is frequently observed in publications. We believe it is crucial to remember that the initial steps of reading books and collecting field data cannot be replaced by any means. Furthermore, in the realm of Roman history and archaeology, without a comprehensive and detailed understanding of the context, computer-based work is likely to be futile.

Hence, it is imperative to comprehend both the tool and the historical issue. Just as we cannot optimize a computer's performance without understanding its internal workings, proficiency in utilizing a word processor alone does not suffice for coherent composition without writing skills and language knowledge. This analogy applies here as well. Moreover, we must acknowledge another challenge: the knowledge gap hampers the tool or inquiry's full utilization potential.

It is common for readers to encounter colleagues or students presenting GIS analyses with intricate tools and impressive visualizations. However, the fundamental question often arises: What does it signify? What is the research question? What methodology was employed? How were the criteria measured? While creating visually appealing maps using GIS is a legitimate and prevalent practice, its potential surpasses mere visualization. Through examination of information sources and correlation with contemporary geography, GIS facilitates the creation of historical models.

Historical texts comprise the writings left by our predecessors and material records, whether archaeological or otherwise. Particularly in ancient history, much of the documentation we rely on is associated with the past.

If we conceptualize this as a research process aimed at uncovering something new, literature serves as a record of events, while material culture offers specific evidence. Both aspects complement each other and synergize well, especially in formulating research questions. The research process follows a methodology akin to that of any other scientific discipline or field, but the challenge lies in grappling with the specific sets of sources inherent to history. These sources necessitate a highly tailored methodology, and their reliability is subject to debate.

Texts can often be subject to debate, particularly when the intentions behind them are not entirely clear. For example, if Pliny the Elder mentions a specific city, a misinterpretation or ambiguity in his intentions could significantly affect our comprehension. We can engage in debates regarding the sources he utilized, the descriptions he provides, and even the chronology. This scrutiny extends to inquiries of authorship and the motives of individual authors, a frequent occurrence in works discussing events such as civil wars.

However, in certain instances, when we encounter a concrete record indicating the presence of something at a particular moment, and we can verify its context without ambiguity or deliberate alteration, we can attribute it a certain degree of reliability. This raises questions regarding whether we possess access to all facets of the historical mindset or if certain elements are less crucial for understanding the historical inquiry at hand.

In historical research, specific data points can sometimes be overshadowed by broader concepts. The endeavour to collect a multitude of comprehensive data points related to an event, akin to the goals of historians and archaeologists, necessitates the utilization of new data analysis tools. The adoption of GIS represents a qualitative and quantitative leap in our comprehension. However, comprehending their potential impact can be challenging, as it pertains not only to data quality but also to data quantity, akin to literary sources where an excess of data may render them ineffective.

The process of converting this data into a computerized format is of immense importance, and we are continuously adapting to optimize its potential. Emphasizing an understanding of the underlying motivations is paramount. Our efficacy in employing these spatial models hinges on aligning them with the appropriate historical mindset. If we adhere too strictly to a classical perspective of historical and archaeological concepts, we risk overlooking valuable opportunities.

We can transport ourselves back to historical moments where critical perspectives, exemplified by the works of Duncan-Jones, revolutionized certain processes by quantifying and comprehending socioeconomic actions (1974). This capability has been facilitated by our ability to analyse and quantify data. The complexity of spatial models is heightened as we endeavour to understand the functioning of specific territories or landscapes and develop theoretical models that can yield new insights. This induction process operates similarly to methodologies employed in other scientific disciplines.

In conclusion, it is crucial to recognize that while our analytical capacity extends slightly beyond the human mind itself, it remains inherently limited in terms of quantifying data. Therefore, the ability to organize space and places and utilize tools that simulate certain behaviours plays a pivotal role in human actions, such as traversing territories, observing objects, or transporting specific products. Anything that can be quantified becomes subject to analysis.

2.2 Varied Levels of GIS Utilization

In the realm of humanities, the predominant use of GIS centres around its role as a geo-referenced database. In the context of heritage management and preservation, having a comprehensive list of locations or points included in an information layer is indispensable. Such data is vital for conducting effective research, facilitating the correlation of archaeological data with spatial information, and ensuring the integration of new information for future endeavours. Archaeology, in particular, heavily relies on GIS, especially for cataloguing heritage assets. The applications are diverse, ranging from simple lists of archaeological sites in a municipality (which eventually contribute to archaeological charts) to projects focused on the study and documentation of various elements, including intangible culture, industrial heritage, architectural heritage, and pure archaeological heritage (Pardo 2008; Sánchez-Aparicio et al. 2020).

However, the potential of GIS transcends these basic uses, particularly from a research perspective. Generally, the depth of technical knowledge required determines the spectrum of applications. In this regard, we can classify GIS usage into three levels based on proficiency in GIS knowledge. This classification is not intended negatively; rather, it acknowledges that accessing its full potential requires a progressively higher level of technical expertise.

Initial Knowledge Level: This stage involves using GIS to generate thematic maps and serve as a form of cartographic representation, often encountered by beginners. While prevalent in scientific publications, caution is warranted. It is crucial to accompany the represented data with explanations and descriptions of the processes employed. Simply drawing a line on a map and declaring it as the ideal route from one place to another does not constitute a rigorous GIS analysis. Vigilance is necessary in this regard, as a significant portion of literature employing this type of analysis may be misleading. Advocates of simplification often advocate for using basic tools without substantial background knowledge, seeking quick representation. Consequently, the introduced parameters may be erroneous, leading to historical inquiries that are either non-existent or inadequate. This lack of methodological rigour can be easily discerned, conveniently reinforcing underlying principles. While a simple visual representation of an element may not necessitate extensive technical expertise, it is indispensable for meaningful analysis. Nonetheless, this approach remains the most commonly employed, particularly when the objective is solely to visually present information.

Medium Knowledge Level: At this stage, GIS is utilized as a database to store and analyse data, in addition to its role in digital representation. This level of analysis is frequently observed in government administrations and among researchers involved in site cataloguing. It requires a deeper understanding of working with fields and data manipulation within the software. Additionally, GIS serves as a workspace for integrating different layers of information, akin to a desktop where various data can be stored and visualized together. This involves working with proprietary data, refining it, and incorporating information from third parties. Essentially, GIS is used as a tool akin to a word processor, bringing together information from diverse sources such as photographs, tables, references, and ideas.

Advanced Level: This book focuses on the advanced utilization of GIS. It revolves around the utilization of historical and geographical data through a range of GIS tools, enabling the creation of representations of reality. This advanced usage constitutes what we refer to as a historical model, as further explored in subsequent sections. It entails employing a diverse array of tools, briefly explored in the subsequent chapter. However, the advanced level signifies the point at which the learning curve becomes significantly steep, making it challenging to comprehend all the tools and potentials the software offers. Fortunately, our emphasis lies not so much in mastering the intricate details of these tools, but rather in acquiring the skill of formulating precise inquiries and conducting thorough information searches. These tasks encompass conducting advanced visibility or mobility analyses, transforming source data to generate novel information, creating bespoke digital elevation models, or engaging in theoretical modelling.

In summary, this book does not delve into the basic or intermediate levels of GIS use but instead focuses on the advanced level. It explores the utilization of historical and geographical data through various GIS tools to construct historical models.

The challenge before us is twofold: firstly, we must relinquish certain, fixed and unalterable stereotypes, and secondly, we endeavour to quantify an increasing number of aspects of daily life. Presently, these efforts continuously manifest in various forms of technology surrounding us, operating with remarkable efficiency. For instance, if a transportation enterprise needs to optimize routes and reduce shipping costs, various systems can accomplish this task. It becomes evident that even achieving a modest 10 per cent increase in precision can yield remarkable efficiency improvements, making the endeavour worthwhile.

Primarily, the effectiveness of GIS use hinges on the user's technical knowledge. This underscores the importance of understanding the specific program and its applications. Consequently, we categorize users into different levels of proficiency, with a foundational understanding based on their familiarity with GIS. Essentially, GIS serves as a tool that enables us to comprehend the relationship between space, time, and human desires within societies.

So, how do we establish this relationship? We do so by examining the fundamental human capabilities related to spatial intelligence, including our understanding of our surroundings, our ability to navigate, and our capacity to recall routes and time spent in moving. This perspective encompasses a range of basic human elements within the technical realm, particularly in the context of historical narratives.

However, Geographic Information Systems represent a significant evolution of a much older practice, dating back to the earliest human records of land surveying, cartography, and map-making. This practice has persisted since the dawn of written history, aimed at organizing and delineating territory. It addresses concerns such as tax collection, land ownership, and land use regulations. These historical issues are now translated into computer programs that greatly simplify the process, reducing the need for extensive prior knowledge. This underscores that the focus is not solely on technical proficiency but also on historical and cognitive perspectives, as further explored.

If we lack a specific historical question or the cognitive tools to formulate such a query, our technical knowledge, regardless of its depth, will not enable us to articulate these concepts effectively. It is crucial to acknowledge this, even with a profound understanding of the subject matter. To illustrate this point, we can look to the example of Eratosthenes in the Roman era. Eratosthenes was able to approximate the Earth's radius, highlighting the importance of bridging the gap between our knowledge and the historical context, allowing us to pose relevant questions and uncover valuable insights.

2.3 GIS Models and Historical Models

In our understanding, a model serves as a representation of reality, offering a theoretical approximation of a process. In the realm of historical research, a model aims to aid in understanding and analysing hypotheses using empirical data. An illustrative example of a contemporary model is a flight simulation, which simulates key components of piloting without involving an actual aircraft. All scientific disciplines necessitate the generation of empirical models to predict real-world responses. For instance, meteorological studies forecast future weather conditions. While these models may not be perfect, their accuracy remains crucial. A dependable model exhibits low uncertainty and has been tested with ample empirical data to ensure robustness.

However, historical models are more intricate, as past processes cannot be empirically tested through practical means. The inherent uncertainty in historical models presents challenges for the humanities in general and is often perceived as a weakness of history as a scientific discipline. Nonetheless, the absence of a time machine to verify the accuracy of historical claims does not render the creation of models incapable of addressing inquiries (Brughmans 2013). Similar challenges exist in other scientific fields, such as tectonics or astrophysics, where direct analysis of underground plates or black holes is unfeasible. The disparity lies in the ability to measure and test phenomena with greater precision in those fields, while historical and archaeological studies confront difficulties in establishing robust contrasts.

The intended meaning of this assertion is that the historical method, despite purporting to be scientific, does not yet possess the same level of scientific rigour as pure sciences. This assertion can be attributed to three primary reasons. Firstly, the materials we engage with, aside from a few questions relating to archaeological artefacts or the interpretation of written legacies, often lack interpretive precision. Secondly, the level of precision required in history does not permit us to create models with 99 per cent certainty. However, formulating two or three plausible explanations while eliminating others constitutes significant progress. Lastly, the third reason pertains to the research tradition itself. By comparing the current state of historical inquiry to the process of biological evolution before Darwin, persistent challenges such as the absence of a unified method and a scarcity of well-documented studies are evident.

Addressing these issues proves to be no small feat. In the realm of history, we humbly acknowledge that many of these methods are still in development.

Consequently, a historical model serves as a theoretical approximation we develop to understand the determinants and behaviours of past societies, constituting the essence of research. However, confusion sometimes arises between a GIS model and a historical model, with the terms used interchangeably, leading to misunderstandings. A GIS model represents an analytical application of data, such as modelling communication routes between two cities to determine the shortest path. In this case, the model generates new information: the road network itself.

As previously defined, a historical model combines existing elements (sources, bibliography) with a historical question (objectively) to arrive at the most accurate representation possible of reality, occasionally incorporating a GIS model. Literary information regarding the existence of a communication route through a specific site and a GIS analysis of the optimal route both serve as forms of new data that we incorporate. Naturally, the reliability of the literary source will typically be higher than that of the GIS analysis.

Therefore, it is crucial to maintain focus on the foundations of our research, the historical questions. They serve as the central axis around which our analysis revolves. Hypotheses, objectives, and methodologies are aligned with these questions. While an ideal outcome would provide a clear-cut answer, it is common to encounter several possibilities. In such cases, we choose the most suitable option by generating various historical models and analysing which one best aligns with reality based on available information sources, experience, and comparative studies.

For example, we could employ a visibility tool to identify the areas under control of a given site. If the methodological process is flawed, we are unable to discern which individual is observing the situation under equally intelligible conditions. Understanding whether the observer is human or colossal and whether they are observing a rider or a dragon with equal clarity, becomes contingent upon internal parameters. Consequently, confusion may arise.

We utilize GIS models to generate historical representations, which may or may not involve the use of GIS tools. However, if such tools are employed, it is crucial to consider the mindset and historical reality within the parameters of the tools in use.

There are numerous examples of historical models. The deconstruction of the past to understand its primary components has always had several focal points, ranging from the political history of the warfare of institutions and religions, and to a more Marxist perspective focused on economics. Currently, perspectives lean towards the study of micro-histories, addressing issues related to highlighting marginalized spaces or invisible social and cultural groups. Similarly, there is a growing emphasis on the transferability of the study of the past, with perspectives related to ecology and environmentalism, such that the study of the past serves to understand and address problems and challenges in a resilient manner.

Historical models are applicable to all periods and locations. Their reliability fundamentally depends on the quantity of data available and the need to extrapolate information. For instance, for a question that occurred fifty years ago, the reliability of reproduction is quite certain, such as reconstructing the path taken by a specific army, given its route, along with diaries, photographs, and oral testimonies. However, as we go further back in time, this reliability diminishes, as we must rely on external and less precise sources, rendering it impossible to answer all questions. The clearest example is in prehistory, where the absence of written documentation necessitates the use of broader models based on archaeological sources. This merely determines the difference in scale and precision of our data.

Therefore, when discussing models, there are numerous examples contingent upon the specific questions at hand. Naturally, we focus on aspects with territorial, environmental, or economic perspectives, as these are fundamental aspects readily incorporable into a GIS. However, historical inquiries may also stem from other origins, such as understanding the dissemination of a particular ritual or cult, or discerning the implications of specific legislation in the practical context in which it was implemented.

It is important to clarify that the approach proposed in this book is not divorced from current research trends. Rather, it aims to effectively combine classical viewpoints with more technocratic ones. In this context, it is common to encounter studies where the historical question is well formulated, and researchers seek to address broader questions beyond merely understanding the functioning of a specific site. They delve into aspects like mentality, politics, economy, religion, or society of the time. Through meticulous analysis of literary sources, it is possible to approach a comprehensive understanding. However, the challenge with these sources lies in their limited reproducibility and verifiability. We rely on the credibility of the author providing the data, corroborating auxiliary sources, or the researcher's own expertise and ability to infer data.

Nevertheless, this methodological approach has spawned several robust lines of research, albeit susceptible to transformation or disappearance. This susceptibility arises because the underlying principles guiding a question may be compromised by the researcher's own preconceptions. To illustrate, consider the question of whether technological advancements occur simultaneously in various regions or emanate from a single point to others. In the case of the ancient world, particularly in Europe, there is a prevailing notion that technology and cultural aspects emanate from the East to the West, while concurrently existing in parallel without one being the origin of the other. Answering this intricate question necessitates consideration of a multitude of variables that cannot all be analysed by a single individual or team.

However, a certain precept has emerged, at least in the field of History, wherein the professorship holds a kind of authority. When someone is deemed the foremost expert in a particular subject, their foundational ideas, whether well argued and defended or not, often become a sort of dogma that must be adhered to, with little room for criticism. Additionally, in our discipline, rightly so, we possess the virtue of reviewing the entire research trajectory in pursuit of historiography. This means that we do not automatically discard precepts from the past, allowing us to trace the origins and evolution of ideas. We say virtue because in other fields, the focus is often solely on the latest and most prominent developments, disregarding earlier contributions.

This phenomenon is indeed beginning to manifest with the advent of technological advancements in historical and archaeological disciplines, which yield irrefutable data. In the past, one could question the dating of a material based on other evidence, such as discovered coins. However, with the emergence of precise dating methods, we now have access to an exact timeframe. Without diminishing the importance and progress that historical and archaeological studies have achieved in this regard, it appears that we have transitioned from one challenge to another. We have shifted from a lack of means to verify the precepts of historical questions to possessing methods for validating them, but perhaps at the cost of losing their essence.

Hence, the proposal presented here advocates revisiting classical historical questions based on a profound understanding of various forms of literature, including texts, legislation, and epigraphy. The objective is to develop technical models to address these questions, thereby enabling us to tackle inquiries that are more complex. A significant advantage of this approach is that our analyses, utilizing Geographic Information Systems, can be easily verified and reproduced by any other researcher. Consequently, if a result is demonstrated, it does not have to be accepted solely based on the reputation or capability of the researcher, or even the prevailing historiographical perspective. The outcome can be entirely objective, immune to human errors in this regard.

However, the current limitations of these historical questions of models lie in two complementary areas. Firstly, we need to translate literary knowledge (when we have references) in a manner that often necessitates quantifying qualitative data. Currently, the introduction of values relies on a process of trial and error. This limitation aligns with the second challenge, which is the scarcity of specific studies on this topic, not only to comprehend individual case studies but also their reproducibility in a broader context. Both of these challenges are expected to be addressed and improved upon in the future.

For history and virtually all areas of archaeology, GIS serves as a valuable working platform. However, a slight distinction can be made regarding their usage in both cases. From our perspective, the models we create are not solely focused on archaeological studies, even if they are built upon archaeological data. Rather, they are historical models. While these models can be applied independently in both fields, the fundamental focus of the study is historical in nature, aiming to answer questions about past societies, economies, or cultures.

We believe it is necessary to metaphorically break down the boundaries between the traditional approaches of historians, focused on written sources, and archaeologists, solely interested in archaeological materials. To work with this type of historical model, there needs to be a harmonious integration of both sources of information and a prioritization of historical questions. This framework allows for the exploration of additional archaeological inquiries within the broader historical context.

2.4 Importance of the Historical Question

We emphasize the importance of the mindset as a central element for understanding historical models, alongside the historical question itself. To reconstruct or simulate historical realities, we must begin by understanding the mindset of the period under study. However, what is the reality we seek to comprehend?

Typically, these approaches relate to questions concerning the landscape, territory, or economy, such as modelling site density, analysing visibility or mobility, and studying territorial boundaries. However, our focus in this book is not solely on these aspects, but rather on models that simulate specific realities and aim to answer specific spatial questions. Rather than conducting a mobility analysis between two cities to create a corridor, our aim is to address the question: How can we best understand human mobility in this area? In other words, it is not about exploring the potential of GIS for our study, but about formulating a historical question and then employing GIS tools to simulate an answer.

Therefore, the historical question must always be at the forefront of our analysis. It should be accompanied by a deep understanding of the mindset of the historical period being studied. Without this understanding, we risk seeking answers to the wrong questions. For example, if we want to estimate the market cost of a particular product but fail to consider that in the ancient world, market interchange was only one among many economic forms and that products were regulated or exchanged within subsistence or prestige circuits, then the model we seek will not correspond to reality.

We must be cautious of poor practices that can be observed in GIS, as it is a quick and easy way to produce scientific material. However, due to its technical nature, it can also lead to errors or internal inconsistencies. Such practices can invite justifiable criticisms, as they may involve approximations that lack real data and therefore hold no historical value in themselves. It is important to be vigilant in identifying the following issues:

Scientific works that aim for sensationalism, presenting visually striking images and drawing risky conclusions, but without a methodological section to support their claims, pose a significant concern. It is crucial to recognize that any image generated with GIS is simply a representation, a still photo of a much broader process. Thus, the focus should be on the accompanying methodological description, and if this is absent or vague, the work may be questionable. Without a clear methodological framework, it becomes challenging for readers to evaluate the validity and reliability of the findings presented in such works.

Conversely, there may be works with excessively complex methods, numerous references, and extensive bibliographies to support them, but which are relatively incomprehensible to the average reader. In such cases, the reader lacks the tools to assess whether the process described is valid or not. Even if the method is indeed valid, this practice is problematic as it hinders reproducibility. Unlike the development of a new analytical technique, reproducibility in this context should also involve comprehensibility. Therefore, it is essential for researchers to strike a balance between providing sufficient detail to ensure transparency and accessibility of their methods.

Theoretical approaches that lack well-tested historical realities and instead use GIS to demonstrate a specific hypothesis are another concern. In such cases, researchers may generate analyses that fit with their preconceived ideas, potentially leading to biased interpretations. While hypotheses are an essential part of the research process, it is important to acknowledge that they will not always hold true. The issue here lies not solely with the researchers themselves, but with the broader system, as high-impact publications and time constraints often prioritize achieving positive results. By being aware of these potential pitfalls and adhering to rigorous methodologies grounded in empirical evidence, researchers can mitigate the risk of bias and ensure the reliability and validity of their historical models developed using GIS.

The issue of using GIS to create historical models should be approached with caution. It is important to note that GIS is not essential for representing historical reality; rather, it serves as a tool to visualize geographical data that would otherwise be challenging to comprehend. It also enables mathematical calculations to combine this information based on the analysed historical mindset. In essence, it is a transcription of ancient ideas into current geographical criteria to the best extent possible. Researchers must remain vigilant in ensuring that GIS-based historical models are grounded in robust methodologies and are used judiciously to enhance our understanding of the past without compromising accuracy or integrity.

Such models are empirical approximations of reality and may be subject to criticism or revision by purist authors. The complexity of the method employed, with its numerous elements and calculations, means that even a small error (intentional or unintentional) can completely alter the entire process. While some concessions are inevitable within our approaches, as they are ultimately theories, it is crucial to be aware of them, express them clearly to the reader, and justify our work transparently rather than hiding them.

In addition to these challenges, technical proficiency in using GIS is also necessary. To enhance their competence, tools need to be specifically designed in this context. While there is still progress to be made in this regard, this handbook aims to display at least one approach to addressing these challenges. Therefore, we will begin by providing a simple description of each of the tools involved in the technical processes, allowing for the reproducibility of the models described in subsequent chapters.

Indeed, historical models developed using GIS are empirical approximations of reality and may face criticism or revision by purist authors. The complexity of the method employed, with its numerous elements and calculations, means that even a small error, whether intentional or unintentional, can significantly affect the entire process. While some concessions are inevitable within our approaches, as they are ultimately theories, it is crucial to be aware of them, express them clearly to the reader, and justify our work transparently rather than hiding them.

3

GIS Software and Tips

In this chapter, we embark on a concise exploration of the utilization of Geographic Information Systems (GIS) as a tool. We begin by outlining the indisputable principles of GIS in a concise manner and gradually delve into more intricate tools, such as the examination of social networks, which may prove beneficial to the reader. Subsequent chapters will individually analyse each model, hence our focus here is to provide elucidations pertaining to the fundamental tools and processes. These explanations are essential for comprehending forthcoming activities. Notably, we aim to offer valuable tips and techniques for their application, while also explicitly delineating the tools and challenges encountered in specific foundational processes. This approach enables readers to revisit this section should any uncertainties arise or if they encounter difficulty understanding particular processes in subsequent sections.

3.1 GIS Principles

In terms of GIS software, two prominent options are ESRI's ArcGIS and ArcPro, as well as the free and open-source software QGIS. Both offer significant potential, but they have different strengths and weaknesses. ArcPro has undergone extensive development and refinement, resulting in polished processes. ArcGIS is the previous version of ArcPro, which is also highly polished and widely used but is now discontinued in updates. On the other hand, QGIS provides a high level of functionality, incorporating open-source code from various developers, but it may be less stable due to frequent updates and changes. In this book, we will describe the processes and tools without favouring one software over the other, mentioning specifically which tool in each software allows us to perform a particular process.

When starting to work with GIS, it is essential to understand the workspace, the types of elements that can be created and modified, the integration of information, and the available tools. This knowledge is relevant for beginners, and experienced users can refer to the list of tools provided at the end of this section to see if any are useful. In the following sections, we will develop and explain the tools that will be used, specifying in the table which tool in each software allows us to perform a specific process. The question here is not to get into the general use of the tool, but how to adapt it to our specific interests, as we will see below.

For newcomers to GIS, the abundance of buttons on the screen can be overwhelming, similar to other software applications like 3D processing, image editing, or sound editing. Many users simply want to create a basic map with points but may find it daunting due to the multitude of tools. As a result, they may resort to using simpler options like Google Earth. However, there is no need to worry. The interface of any GIS typically consists of a main screen where information is shown, a list on the left displaying each layer or element to be represented, and a series of buttons above. In practice, we will only use a few of these buttons, such as those for creating or editing data. The remaining buttons will generally not be necessary for our purposes, and the others will be described in subsequent sections.

When it comes to representing reality in GIS, we have two primary methods: entities and bitmaps (vectors and rasters). All other representations are variations of these two. Entities can be classified into three types. The simplest is a point, which represents an element or entity with X and Y coordinates associated with a data table containing additional information such as names, categories, or altitudes. Connecting multiple points creates the second type of entity: lines. When lines converge and close, they form polygons, which are the third type of entity. Despite the potential complexity of these shapes, any element we include in our database can be categorized into these three types.

While this may be a generalization for individuals with more advanced knowledge, it is crucial not to overlook the fact that all non-raster information can be classified into one of three categories. These categories essentially represent the fundamental forms in which objects in space can be conceptualized. Establishing the geometry type is fundamental for commencing a study. For instance, consider a scenario where we aim to analyse and compare a Roman road with the surrounding soils it traverses. We possess a line representing the road and multiple polygons representing the soil types. How can we establish their relationship? Since the line lacks width information, our only option is to intersect it with the polygons. However, by utilizing basic tools like the buffer function, we can create a zone of influence on either side of the line, converting it into a polygon. This allows for interpolation with the soil polygons. This simple case study effectively illustrates a fundamental challenge: comprehending these three categories and the necessity to convert data into the same category to enable meaningful comparisons.

The other representation method is a raster, which consists of a grid of pixels or coloured dots forming a matrix. It is analogous to a photograph where each cell represents a colour. Rasters can be used to depict cartography or other types of information such as terrain models, as we will explore later.

Processing entities (points, lines, polygons) and rasters used different tools and methods. When working with entities, whether individual points or connected polygons, the focus is on transformations such as joining, separating, or trimming. These operations may seem straightforward in theory but can become more complex in practice. For instance, transforming a set of points into a mesh or topographic data for terrain modelling requires tools that convert points into rasters.

On a basic level, there are four main types of GIS tools. The first type consists of simple tools that modify or edit specific elements. The second type includes tools that interact with each other, such as clipping or summing, to perform operations on multiple elements simultaneously. The third type is transformation tools, which change the format or representation of data, such as converting polygons into rasters. Lastly, spatial analyst tools are more complex and combine multiple tools and algorithms to perform advanced spatial analysis.

Fortunately, most of these processes are straightforward to execute. The GIS software includes built-in tools, and users only need to know which tool to use and under which circumstances. For detailed instructions, users can refer to the program's manual or introductory GIS books like Bolstad (2003).

When working with GIS data, it is important to consider that the information may come in different formats and coordinate systems. Transforming data from one format to another is usually relatively easy, as long as the original coordinates are known. However, problems may arise when the data is in a different coordinate system or projection.

There are several factors to consider when dealing with coordinate systems. Firstly, the type of coordinate system can be geographic or projected. Secondly, the data may be associated with a specific zone or region, similar to time zones, where each one is assigned a specific number. For example, Great Britain and central Spain are in zone 30, while Portugal and Ireland are in zone 29. Projection issues can also arise, as different projections use different models to project the Earth's curvature onto a two-dimensional surface. However, these issues are not commonly encountered in most cases. In Europe, a general reference system called ETRS89 has been in use for several years.

The choice of reference system depends on the scale of the survey. For large areas encompassing multiple countries, it is advisable to use a general reference system consistently. On the other hand, for specific projects focused on a small area, using the reference system specific to that area can help avoid transformation problems.

When working with data, particularly in archaeology, it is important to exercise caution as not all public administrations may have up-to-date information. Outdated data or data in old coordinate systems can pose challenges. In such cases, it is crucial to study and understand the coordinate transformation process. Using reference points of known archaeological sites can help ensure the accurate placement of data after transformation.

In summary, understanding coordinate systems, projections, and data formats is essential when working with GIS. Being aware of potential issues and verifying data accuracy through control points can help mitigate any problems that may arise. The resume of each set, along with their usefulness and their location in ArcPro and QGIS is given in Table 1.

3.2 Useful Tools and Tips

The average user tends to find working with points, lines, and polygons easier than raster formats, mainly because they are simpler to understand and manipulate. For instance, in a typical scenario, we may have an information layer containing different points and polygons representing archaeological

Table 1 Summary table of the tools developed below, with an indication of their location in ArcPro and QGIS

Process	Tool	Description	ArcPro	QGIS
Influence area	Buffer	Influence parallel area	Analysis tools / Proximity / Buffer	Vector / Geoprocessing tools / Buffer
Extract	Clip	Cut by mask	Analysis tools / Extract / Clip	Vector / Geoprocessing tools / Clip
	Split	Divide by mask	Analysis tools / Extract / Split	Advanced Digitising / Split feature
Union	Union	Merge multiple layers	Analysis tools / Overlay / Union	Vector / Geoprocessing tools / Union
Dissolve	Dissolve	Union of entities with the same field	Data Management tools / Generalisation / Dissolve	Vector / Geoprocessing tools / Dissolve
Average distances	Thiessen or Voronoi polygons	Lines with an average distance between entities	Analysis tools / Proximity / Create Thiessen polygons	Vector / Geometry tools / Voronoi polygons
Field Calculator		Calculate field information	right click on the field	right click on the field
Slope	Slope	Slope in degrees or percentage	Spatial analyst tools/ Surface / Slope	Raster / Analysis / Slope
Orientation	Aspect	Orientation to cardinal points	Spatial analyst tools/ Surface / Aspect	Raster/ Analysis/ Aspect
Insolation	Solar radiation	Index of solar radiation	Spatial analyst tools/ Solar radiation/ Area solar radiation	GRASS/ Raster / r.sun. insoltime
Visibility	Viewshed	Visual and non-visual areas from an entity	Spatial analyst tools/ Surface / Viewshed	GRASS/ Raster / Viewshed
Mobility	Path distance	Cost raster from a given entity	Spatial analyst tools/ Distance / Path distance	GRASS/ Raster / r.cost
	Cost path	Less cost path from the previous process	Spatial analyst tools/ Distance / Cost path	GRASS/ Raster / r.walk

(Continued)

Table 1 Continued

Process	Tool	Description	ArcPro	QGIS
Hydrology	Flow direction	The direction of water flow	Spatial analyst tools/ Hydrology/ Flow direction	GRASS/ Raster / r.drain
	Flow accumulation	Accumulation of water flow	Spatial analyst tools/ Hydrology/ Flow accumulation	GRASS/ Raster / r.flow
	Watershed	Area of superficial water convergence	Spatial analyst tools/ Hydrology/ Watershed	GRASS/ Raster / r.watershed
Topographic profile	Terrain profile	Cut of terrain in Z	3d Analyst / Terrain graph	Terrain profile tool
Coordinates	Projections and transformations tools	Change coordinate system and projections	Data management tools / Projections and transformations	In the properties of the layer
Transform	From raster	Transform from raster to polygons, lines or points	Conversion tools / From raster	Raster / Conversion / Polygonise
	To raster	Transform from polygons, lines or points to raster	Conversion tools / To raster	Raster / Conversion / Rasterise
Extraction	Multiple tools	Extract one raster with another or a polygon	Spatial analyst tools/ Extraction	Raster/ Extraction
Interpolation	Multiple tools	Fill the gaps between points, depending on the tool itself	Spatial analyst tools/ Interpolation	Raster/ Interpolation
Reclassify	Reclassify	Modify the values of a raster	Spatial Analyst tools/ Reclass/ Reclassify	GRASS/ Raster / r.reclass
Raster calculator	Raster calculator	Modify one or more raster values	Spatial Analyst tools/ Map algebra/ Raster calculator	Raster/ Raster calculator
Processing models	The model builder or Process toolbox	Interconnect multiple tools	In the same main bar	Process toolbox/ Create a new model
Program codes	Python console	Modify tools and create new processes manually	in the same main bar	Plugins/ Python console

protection zones where a construction project is planned. Tools like influence area, union, dissolve, or extract can be used to modify and create new information within these layers. These tools are relatively straightforward to use and do not require extensive explanation.

We consider these to be relatively straightforward concepts, such as intersecting one line with another or combining multiple elements. In the case of raster data, the main challenge lies in the type of information it represents and the fact that transforming it alters the original data. For instance, when we have a polygon representing a neighbourhood grid, we understand that it signifies a specific block. Regardless of whether we cut or add to it, this unit remains the block, even if it is divided by a street. In contrast, with raster data, the value assigned to each grid cell represents a particular attribute, such as elevation above sea level. Modifying raster data entails altering this value. For example, we may remove coastal or water areas, thereby altering the data. However, modifying it in this context raises the question: what does the new value represent? If we were to combine a slope map with an elevation map, the resulting value would be nonsensical, lacking any meaningful interpretation.

However, it is important to highlight the use of field calculations in entities, which allow us to add and modify information within the attribute fields of the data. This can be useful for adding new data to our database and for the modelling process, as we may need to extract values from previous datasets to incorporate into the final output. While this tool may be more complex for users with limited GIS knowledge, we will provide tips and guidance for each specific model that involves its usage.

In archaeology, other commonly used tools include natural neighbour interpolation and kernel analysis, which are employed to calculate artefact or site density (Sibson 1981; Yin 2020). Such tools can also be used to model historical issues, in that they allow us to know the relationship of particular places to each other or to other elements. In general terms, it is a matter of converting punctual information into raster, based on visually filling in the limits between them. We can also extract numerical data from connections, such as nearest neighbour. For example, in the case of having several Roman villas and a layer of rivers and streams, we can determine which one is the closest to each of the elements.

When working with raster data, it is important to consider that we may need to obtain data from external sources. While point and polygon data can often be collected directly from fieldwork, one of the key datasets required for most analyses is topographic information of the terrain, typically represented through a terrain model. There are three main types of terrain models: DEM (Digital Elevation Model), DSM (Digital Surface Model), and DTM (Digital Terrain Model). DEM represents the natural surface without any alterations; DSM includes all surface features such as buildings, rivers, and forests, while DTM focuses on the natural terrain modifications, excluding vegetation and buildings. A DSM can be interpolated to generate a DTM, but the reverse is not possible. The accessible data can be in processed or raw formats. If it is already processed, we must work with the available data. However, if it is in raw format, we have the opportunity to filter and enhance the resolution and detail of the data in some cases (El-Sheimy et al. 2005). Understanding the nature of the data we are working with is of utmost importance. When dealing with unfiltered data, it encompasses both anthropic and natural elements, which can introduce errors if we treat it as a true representation of the Earth's surface. Similarly, when using pre-filtered data, there may be minor imperfections or errors that result in false positives during our analysis. Hence, it is crucial to ascertain the type of data being utilized and perform manual checks. Typically, importing a small area and comparing it against an accurate ortho-photo in a disrupted region, such as a recently developed rural settlement or a bridge, suffices to observe how the data has been transformed. This verification process allows us to identify any discrepancies or distortions.

Terrain data can be obtained from two main sources. Photogrammetry or Light Detection and Ranging (LiDAR) laser scanning data. In addition, it can be collected by us for small areas or obtained through national or international agencies for larger areas. Aerial photogrammetry is commonly used for small, high-resolution areas surveyed by drones, or for larger regions through aerial flights or satellite imagery. LiDAR is another special type of laser technology used to measure the surface. LiDAR applications in archaeology are well known, as they can generate highly precise terrain models, allowing for the detection of archaeological sites (Chase et al. 2017; Monterroso 2017). While we will not delve into the details of LiDAR processing in this book, it is important to note that it also requires GIS processing. The data generated by

LiDAR surveys provides an ultra-precise DTM that can be used for archaeological site detection and analysis (Guyot et al. 2018; Zani and de Rosa 2018).

In aerial photogrammetry, the process involves constructing photogrammetric models by superimposing images with a large overlap to determine the three-dimensional coordinates. LiDAR, on the other hand, utilizes laser technology to measure the Earth's surface with high precision, even in challenging areas with low or dense vegetation. The density of LiDAR points allows for the classification and removal of anthropogenic or natural disturbances like buildings and forest stands, making it increasingly popular for generating accurate digital terrain models with pixel densities easily less than 0.5 metres.

When working with such data, several considerations must be taken into account. Firstly, it is important to differentiate between working with a digital elevation model (DEM) or a digital surface model (DSM). If the model has already been filtered, the removal of anthropogenic elements like buildings and roads must be confirmed. This is important if we work with LiDAR data. If no filtering has been applied, additional data processing may be required. Including anthropogenic elements in the analysis can be problematic, particularly in urban or densely populated areas. The processes for correctly removing those elements could be hard to summarize, as it depends deeply on the type of element and region (Poirier 2013; Cerrillo and López 2020; Lozis and Stular 2021).

The scale of the study area constitutes another fundamental factor to consider. Scale, in this context, denotes the magnitude of the study area, wherein a larger scale signifies a smaller study area, and conversely, a smaller scale signifies a larger study area. The resolution of raster data is contingent upon the chosen scale. For the meticulous identification of archaeological features, the utilization of high-resolution data is imperative. LiDAR surveys have the capability to attain pixel sizes as diminutive as 0.5 metres or even smaller, whereas aerial photogrammetry or LiDAR conducted via drones can furnish accuracy at the centimetre level. Nonetheless, due to practical limitations and resource constraints, the majority of researchers must contend with raster models that possess certain limitations concerning scale.

It is paramount to strike a delicate equilibrium between precision and computational capacity when determining the optimal scale for the study area.

Overly high precision may engender datasets teeming with billions of data points, thereby substantially elongating processing durations or surpassing the computational capabilities of the hardware. Analogous to the precision requirements of GPS, wherein micron-level accuracy is unnecessary for measuring a building, the same principle applies to model resolution. A model featuring a resolution of 5 metres may yield outcomes akin to those derived from a model with a resolution of 0.5 metres, albeit with markedly diminished processing durations.

The requisite precision and resolution of the model are contingent upon the unique characteristics of the study area and the objectives of the research. For a singular archaeological site and its immediate environs, achieving high precision is imperative. However, when addressing broader scales, such as multiple sites encompassing a city's jurisdiction, a resolution ranging from 1 to 5 metres may prove adequate. In more expansive regional contexts, a resolution spanning from 5 to 25 metres could be deemed acceptable, whereas analyses conducted on an international scale might necessitate resolutions surpassing 25 metres. These examples are illustrative, and the optimal resolution will hinge upon the specific circumstances of the study. It is crucial to factor in the computational resources required, as heightened precision and resolution typically demand greater computational capacity and processing time.

Naturally, these considerations will fluctuate contingent on the locale, given that each country and region may possess distinct data accessibility. In certain instances, copious and readily available information might be accessible, whereas in others, it may be restricted, necessitate remuneration, or remain entirely inaccessible. Consequently, conducting certain spatial analyses might be impeded by constraints stemming from data resolution. In scenarios where the desired resolution is unattainable, the resultant model may lack validity or produce disparate outcomes compared to utilizing alternative data sources. For example, notable disparities in data resolution between Spain and Portugal have been personally observed, notwithstanding the political demarcation between the two countries, which does not coincide with a tangible geographic boundary dating back to Roman times. Further exploration of this issue will be undertaken in the subsequent chapter, which delves into resource accessibility.

Nevertheless, raster formats, as elucidated previously, are particularly instrumental for terrain mapping comprehension, with an array of tools at our disposal for transforming or modifying general data into spatial transformations. Numerous tools are available for this purpose, encompassing extraction and interpolation tools. Additionally, the consideration of coordinate and transformation tools is crucial, as data sourced externally may necessitate adjustments to align with our coordinate system or attain the desired resolution. Transforming the raster, such as generating a mosaic featuring varying resolutions, assumes particular relevance when grappling with expansive areas. While a high resolution, such as 5 x 5 metres, may confer advantages for smaller regions, the projection of the raster to a broader resolution, such as 25 x 25 metres, may become imperative when addressing a nationwide scale. The determinations regarding these transformations will hinge upon the computational capabilities at our disposal and the precise scope of the project.

3.3 Advanced Process

Whenever we are interested in spatial planning or environmental issues, GIS provides easily accessible and user-friendly tools that have been extensively tested in these applications. However, certain raster-related tasks may require more knowledge and understanding of tools such as reclassification, raster calculator, or combinations of tools. This is particularly pertinent when employing historical models. Therefore, we allocate a dedicated section to these tools as they serve as the technical foundation upon which the subsequent models are built. Broadly speaking, the GIS tools we will employ consist of specific spatial analysis tools, which will be elaborated upon in each respective section. Additionally, we will discuss a set of general tools for reclassifying and integrating data in the following section.

A raster is a grid where each pixel represents a specific value. Reclassification involves selecting ranges of these values and transforming them into different values. To illustrate, imagine taking an image and instructing the software to reclassify it so that all red areas become green. In other words, all pixels that had a red value (e.g., represented by the number 5) will now be assigned a different value, such as the colour green.

Although the meaning behind the numerical values in a spatial representation such as a raster may appear obscure at first glance, it is imperative to grasp the significance of each raster and its underlying units of measurement. Understanding the unit of measurement and the context behind the values is paramount. For instance, in a topographic raster, the values may denote terrain elevations in metres, while in a slope raster, they could represent slope angles in degrees or percentages, depending on the chosen measurement. Reclassification in such instances entails grouping slope values into distinct ranges, such as assigning a new value to slopes spanning every 10 degrees.

Reclassification serves as a means to metamorphose data into an entirely different form. However, it is crucial to acknowledge that numerical values invariably underpin these transformations, even when entities with textual information are being manipulated. Consequently, the outcome of reclassification will invariably yield an integer value.

Comprehending this concept is pivotal, as it often poses a significant challenge for many individuals. For instance, we can categorize different geological types by assigning new values, thereby delineating where each specific soil type occurs within the new raster. Reclassification can be executed based on a rating system, where a value of 10 signifies geological types well-suited for our intended purpose, gradually decreasing to 0. The resultant data layer will comprise scores amalgamating various geological types in accordance with our chosen classification. We will explore specific applications of reclassification within each model.

In addition to reclassification, the raster calculator serves as a versatile and potent tool facilitating numerical calculations involving raster values, application of formulas, or amalgamation of multiple layers. Among the most commonly utilized functions are addition, subtraction, multiplication, and division. For instance, when performing addition, the values of each pixel in two input rasters are summed. It is imperative to ensure that the two rasters possess identical resolutions and to comprehend the nature of the values being merged. Maintaining accuracy necessitates avoiding the mingling of disparate units of measurement. For instance, if we merge slope data with altimetry, where one represents degrees and the other represents metres, the value 3 could signify either 1 metre of elevation and 2 degrees of slope or 2 metres of elevation and 1 degree of slope. As we incorporate more layers into our analysis, these considerations become increasingly intricate.

The range of possibilities is too extensive to explain comprehensively, but what is crucial to understand is that when using the raster calculator, it is necessary to modify the underlying data beforehand. To illustrate this, let us consider a clear example of modifying a digital elevation model by creating a hole in an area where a hill has been excavated by a quarry. In this case, if we have topographic information that represents the previous topography, we can generate a raster with that information. By using reclassification, we can assign the actual elevation value to the hill and a value of zero to the remaining study area. Then, we can crop the original topographic raster only in the quarry area and assign a value of zero to that region. In that case, we have a complete topographic raster with 0 values in the quarry and a new raster with altimetry modifications in the quarry and 0 values in the rest. With the field calculator, we can add both rasters together. As a result, the real topography will be preserved in areas outside the quarry, while the quarry area will have the new data.

As shown in this example, the raster calculator allows us to combine multiple layers of information into one, which is particularly useful when creating models. Interestingly, we can easily aggregate multiple criteria into a single value. For instance, if we have classified several layers using a numerical system, with 0 indicating non-compliance and 10 representing complete compliance, along with intermediate values, we can simply add all the layers using the calculator. Therefore, if there are five layers, a value of 50 would indicate that all the requirements are fully met in those cells.

It is worth noting that sometimes we may want to combine criteria that require whole numbers, as decimals are not allowed in their representation. In such cases, a useful tip for solving it is to multiply the decimals until they become whole numbers and, when the process is finished, then divide them again for the same value to have the original measure.

Another scenario arises when we want to visualize the simultaneous fulfilment of multiple criteria in our model, rather than their sum. This will be explored further when analysing the landscape. In this case, a useful technique is to avoid directly combining the criteria themselves. For example, suppose we have three reclassified layers with values 0 and 1, indicating the presence or absence of a specific condition in each layer. If we added these layers together, we would obtain values of 1, 2, and 3. However, it would not be clear which value corresponds to the fulfilment of each criterion. To address this, a simple approach is to assign different place values such as units, tens, hundreds, and so on. The first raster would have values of (1, 2, 3), the second raster would have values in tens (10, 20, 30), and the third raster would have values in hundreds (100, 200, 300). By adding these rasters, we can determine the fulfilment of multiple criteria. For example, a value of 333 indicates that all three requirements are met, while a value of 313 indicates fulfilment of the first and third criteria but not the second. This method enables the generation of complex models without requiring overly complicated processes in practice.

As you can see with the reclassify and raster calculator, the possibilities are endless. However, the reader will realize that we are working with many different tools, some of which have to be used several times in the same calculation or in parallel for each analysis. As we engage in various processes within GIS, involving different tools and layers, whether they are points, lines, polygons, or rasters, we often need to perform a chain of multiple tasks. For example, when conducting a mobility analysis, we typically need to generate a slope map from a digital elevation model and then modify it to incorporate factors of interest, such as the difficulty of crossing rivers or other natural obstacles. We combine these data to create a cost map, distance map, and backlink map and ultimately generate a route map, which may need to be transformed into a line entity for better visualization. This process may involve using around 10 different tools, which, although individually straightforward, can become tedious and complex to reproduce.

To address these challenges, GIS provides tools such as the Model Builder in ArcPro. The primary advantage of using this tool is that it allows us to automate processes by connecting various tools together. This saves time and enables the use of a set of tools within a unified workflow. However, it is essential to test the use of these tools and have a clear understanding of the manual process before attempting automation.

Sometimes, it is not feasible to encapsulate the entire process within a single workflow. Due to the complexity of certain models, it may be necessary to create multiple process packages by combining several tools to achieve specific results, and then utilize those results as inputs for subsequent processes. This approach is influenced by the desired deliverables and the need to work with intermediate data separately.

Moreover, these tools are particularly useful when working with multiple datasets. For example, if we want to conduct a cost analysis from various archaeological sites, we will need to analyse each site individually. If we have 50 sites, this will require performing an isochrones analysis for each site based on the cost of displacement. In such cases, the model builders offer iterators, which are functions that allow the replication of a set of actions. For instance, we can select a specific number of points from a layer and repeat the process for each point using iterators. It is important to save the results of these iterators independently. In ArcPro, for example, this can be achieved by adding a code such as '%value%' to the name of the output file or any intermediate process that we wish to save. It is worth noting that the remaining layers generated during the process are temporary and overwritten, which eliminates the need to manually move and delete intermediate layers. However, it can be problematic if we intend to use data from intermediate stages of the process.

With the use of iterators, the range of analysis possibilities expands significantly, enabling the creation of new tools. When we build a model using the Model Builder, we can save it as a self-contained tool, allowing for easy reproduction. It is even possible to generate a process package and integrate it into the Model Builder as another tool.

In the chapters dedicated to specific applications, we will delve into the process of constructing these interconnected process packages. While they are also considered tool models, we will refer to them as tool or process packages to differentiate them from historical models. It is important to note that these packages are not always confined to a single closed process, as they often involve manually adjustable parameters, location-specific conditions, or criteria that may be relevant in some cases but not in others. This is why we need to rely on our historical knowledge because most of the models will be related to uses and traditions that may vary depending on the time, society and specific context.

However, building these models or packages of interconnected processes is not always a straightforward task, particularly when it comes to constructing historical models. Therefore, the ability to export and share these packages among researchers is highly valuable, allowing others to install and use them as add-ons.

Nevertheless, there may be instances where additional technical expertise is required, particularly in the realm of using programming code, such as Python, in GIS. This necessitates a basic understanding of the programming language. In tools like the field calculator, raster calculator, and Model Builder, we can incorporate such code. Those familiar with programming languages can develop specific tools without difficulty. However, for those of us without extensive programming knowledge, it is possible to employ code fragments to accomplish specific tasks, particularly when they resemble queries or involve simple calculations. Therefore, if necessary, we should not shy away from utilizing snippets of code. Another common use of code is to incorporate external functions, such as trigonometric calculations involving sine, cosine, and tangent, for calculating certain geographical phenomena. Opening the door, even if only slightly, to the realm of coding offers infinite possibilities. In GIS, we can work directly with code, and there are even dedicated software programs that facilitate this process.

3.4 Social Network and Statistical Analysis

In this section, we briefly touch upon network analysis, although it is not covered in detail in the latter case studies. Despite its importance in archaeology, it is not strictly a GIS methodology and requires the use of external tools, which complicates its inclusion in the described method using ArcGIS or QGIS.

Network analysis, as the name suggests, involves studying various elements of interest, such as archaeological sites, as nodes and connecting them with lines based on their relationships. The aim is to visually observe the level of connectivity between these elements and uncover insights that may not be immediately apparent. For example, we can connect archaeological cities based on the trade of specific objects found within them to establish relationships (Brughmans 2010; Knappett 2013).

Although these methods are not new to our field and have roots in statistical and mathematical components applied in other disciplines, such as environmental studies, there are some criticisms when applied to archaeology. One important critique relates to the historical context, as these analyses do not capture individual decisions and interactions and are often based on long-term cultural processes (Collar et al. 2015). Thus, they may not be suitable for understanding specific realities, but they have potential in two main areas: the generation of theoretical models, as presented in this book and the inference of new data based on existing data.

However, it is important to note that this type of analysis relies heavily on real empirical data. Good quality and quantity of data are essential for meaningful results, although predictive models can be proposed. There are numerous examples of their current use and representation in various contexts (Cooper 2015; Rawat 2021).

It is worth mentioning that these tools are relatively separate from GIS calculations and require the use of external tools. There can also be some confusion regarding terminology, as network analysis within GIS often refers to geographical connections between places, commonly used in urban planning for road and infrastructure networks. The choice of specific tools for network analysis varies widely, with a range of free and paid options available in the market. The selection of tools typically depends on the type of analysis being conducted and the specific requirements of the study.

This type of analysis can be approached from the perspective of graph theory, which is a mathematical method that simplifies complex matrices into a more manageable representation. Currently, this approach is not widely used, except in the field of anthropology and related disciplines. However, Social Network Analysis (SNA) has gained significant popularity, particularly in relation to geographical spaces. When discussing network analysis in archaeology or history, it often refers to SNA.

The primary goal of SNA is to compare individuals based on specific parameters, similar to comparing people of a certain age based on their history, health conditions, dietary habits, sports activities, and medication. In archaeology, the focus is on specific archaeological sites or materials. SNA is valuable for understanding social relationships when individuals are included within the analytical framework (Figure 1). It has been extensively employed in studies related to object movement, trade, and the exchange of goods. Comparative analyses between SNA and GIS mobility analyses have been conducted to explore the interconnectedness of social and geographic relationships.

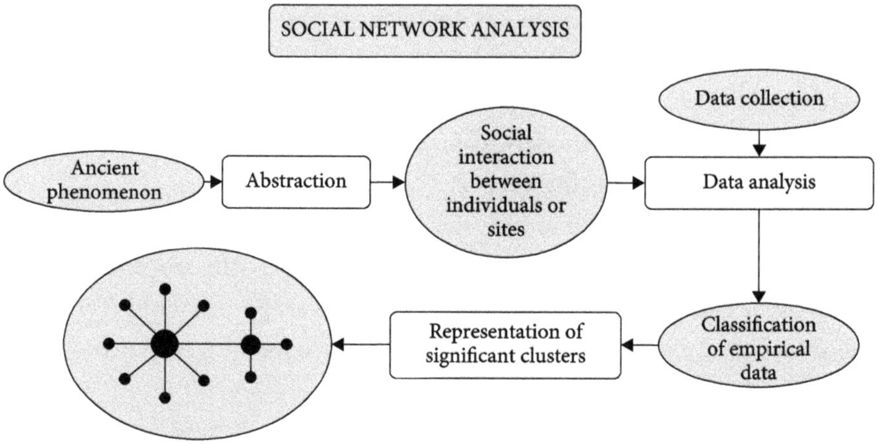

Figure 1 Scheme of social network analysis (source: author).

However, these tools are still undergoing significant development. There is a need to adapt archaeological data to the methods of network analysis and vice versa, enabling the importation of large datasets and the development of specific tools for historical analysis. Additionally, incorporating temporal aspects into the analysis to make it dynamic and not limited to a specific moment would be relevant. Another challenge is establishing the relationship between material culture and real social connections, highlighting the importance of studying historical mentality. Looking ahead, the potential of these tools is expected to be greatly enhanced in the coming years due to increased computing power, the availability of free software (such as UCINET, Pajek, Gephi, Cytoscape, GraphViz, NodeXL, and various packages for the R platform), and the integration of artificial intelligence (Peeples 2019).

In the context of everything we have been discussing, various GIS tools serve to generate analysis models. However, our precepts will always be rooted in our knowledge standpoint. We have already discussed the importance of the historical question, without which none of this makes sense. Now, we add a way to validate our main hypotheses. If the idea is simply to find the best route from one place to another, we can test various optimal routes and choose the most feasible one based on cost or suitability. However, in more complex analyses, quantifying and making sense of the results may be challenging. This could be due to the complexity of the evaluated data, expressed in measures that are difficult to perceive or with large or very small values. For example, estimating whether the cost of 100 hours of land transport is high or low. Moreover, sometimes our results and methodologies may lack verification, particularly when dealing with predictive or other types of processes. The ability to iterate the process to incorporate new or more precise elements into the model is one of the great capabilities of these models.

In both cases, it may be of interest to use tools of probability and statistics to verify the accuracy and validity of the data. This is a complex area beyond the scope of the Humanities, so we will provide a brief overview of the main tools and theoretical precepts that can be applied in a general sense, focusing on those that may be most useful.

Statistics and probability are applied to various scientific aspects, such as predicting climate behaviour. Therefore, there are numerous tools and methodologies available. The most basic and common with any historical period would be cross-comparison, where we examine the historical evolution of a particular issue.

Spatial Distribution Analysis: It utilizes spatial analysis tools to examine the geographic distribution of data. You can look for patterns, clusters, or unusual dispersion that may indicate data accuracy or errors (Mishra et al. 2007). This can be easily done using Kernel Density, which calculates the density of a phenomenon based on the number of nearby points.

There are many other methodologies that can be studied and implemented. Multiple manuals are available on this subject (James et al. 2021). Some of the most useful methodologies and tools are already incorporated into GIS. Below, we describe some of the most important ones:

Spatial Cross-Validation and Interpolation: Apply spatial cross-validation techniques, which are variations of standard cross-validation but take into account the spatial structure of the data. This allows you to evaluate how your model performs in different geographical regions. This can be done through Kriging (Geostatistical Analyst). Similarly, if the data is incomplete, spatial interpolation and extrapolation techniques can be used to estimate values in locations where you have no observed data. In addition to Kriging, you can use Inverse Distance Weighting (Spatial Analyst) to interpolate values using inverse distance weighting.

Spatial Residual Analysis: If you are using regression models in a GIS, you can conduct spatial residual analysis to detect non-random patterns in the residuals that may indicate fitting issues or unconsidered structures in the model. However, in our models, this is likely to be significantly challenging due to the possibility of imprecise variables. This analysis can be performed using Spatial Autocorrelation (Spatial Statistics).

Proximity Analysis: Employ proximity analysis tools to assess how nearby geographic features affect the outcomes. This is useful for verifying whether spatial relationships influence your dependent variables. This precept is particularly relevant in our case, for evaluating the spatial variable concerning assumptions. Most analyses involve a question related to location. However, sometimes the result may be determined by proximity to other elements. This can be quickly accomplished using the Near tool, which calculates the geographic median distance. In addition, complementing this with mobility analysis, as we will see in the corresponding chapter, can facilitate spatial analyses regarding terrain topography. For example, to illustrate its utility, consider conducting a population distribution analysis to understand land parcelling or potential gaps in a territory. While the distribution may seem random based on basic criteria such as proximity to roads or cities, you can launch extensive analyses from central points to other influencing factors, such as soil types or water zones. The statistical value may reveal something significant to correlate a criterion that may initially escape notice.

Spatial Clustering Analysis: Employ spatial clustering analysis techniques to identify data clusters based on geographic proximity. This can help verify if certain areas exhibit different behaviours. This can be performed using tools like Cluster and Outlier Analysis (Spatial Statistics). However, such tools are most useful when a large dataset is available to yield meaningful results.

If our goal is to predict models, we can check the errors in our model by comparing it with real values. This involves testing theoretical assumptions against real data through various methods such as Mean Absolute Error (MAE), Mean Squared Error (MSE), and Relative Mean Squared Error (RMSE) to assess prediction quality.

In this context, regression analysis can be highly valuable (Kleinbaum et al. 2007). It involves analysing regression coefficients, which provide information about how each independent variable influences the dependent variable. You can also analyse p-values to determine if the coefficients are statistically significant. This point can be combined with what is often referred to as hypothesis testing, which evaluates whether the differences between predictions and actual values are statistically significant. Once again, to illustrate with a simple example, we can understand that a vehicle's load is high when it carries 10,000 kilograms instead of 100. However, if our prediction yields values very close to reality but deviates by 10 or 15 per cent, how do we verify their accuracy? We have already mentioned that we cannot expect a high degree of precision in our models. Nevertheless, we cannot simply rely on this and avoid efforts to improve their precision.

Let us focus briefly on explaining one specific case, which is regression analysis. It is especially useful for our cases as it allows us to evaluate the relationship between the dependent variable and one or more independent variables. The primary goal is to understand how independent variables influence the dependent variable and how they can be used to predict its behaviour. In essence, regression analysis seeks to find a functional relationship between variables that best fits the available data. There are several types of regression analysis, but here we will focus on linear regression, one of the most common and fundamental types. Linear regression assumes that the relationship between variables is approximately linear, which is generally what we encounter in our analyses. For example, the dependent variable could be the location of an archaeological site that we want to predict. The independent or predictor variables are the parameters we include in our model (slope, orientation, access to water, etc.). Its application can be somewhat complex, depending on the formula we use.

4

Resources and Methodology

Having reviewed the different tools we can use, we will now define how we access the resources necessary to establish the models and the general methodology for their study. To do this we briefly look at access to the main resources, some of which are of a historical nature such as cartography. Then we dedicate three sections to hydrography, geology and current uses because of the enormous potential they have for the proposed models and because they are external to our discipline. Finally, we end with the general method to be applied to each of the subsequent case studies.

4.1 Access to Resources

When conducting an analysis, it is important to have technical knowledge of the tools mentioned earlier. While there may be other useful tools, the focus in this section is on the aforementioned tools to avoid repeating instructions on how to access and use them. Now, we will adopt a similar approach to discuss different geographical resources that need to be considered, in terms of both access and precautions.

Access to resources primarily depends on the country or region, as it often varies in different locations. In general, most models require at least four fundamental elements: a digital elevation or terrain model, archaeological information, data on the ancient mindset, and current geographical information for comparison. In general terms, the former can be obtained from international organizations, as we will see below. The field and mentality data require our work, the main problem is in the latter group of data, some of which may be operational for the entire public, partially capped or non-existent. In this line, it is advisable to talk to a specialist in the field, so that he/she can tell us what data of edaphological, geological, hydrological, or other, origin we may be overlooking.

The first essential information, which is relatively easy to obtain, is a reconstruction of the topography using a digital elevation or terrain model, as mentioned before. The accuracy of the model is crucial for our purposes, although it also depends on the scale of the study. As previously mentioned, for local studies, scales between 0.5 and 5 metres may be appropriate, while regional scales can range from 5 to 25 metres, and cross-country or international scales typically require 25 metres. These are average values and can vary based on specific project requirements. The scale of the work also corresponds to the scale of the geographical resources used. In general, digital models of around 20 or 25 metres can be suitable for working in larger areas. In this regard, regardless of the country, data from sources such as Copernicus (EU-DEM v1.1. 2022) are available. These data provide 25-metre models in sheets that are relatively easy to access and can be downloaded for free.

During the analysis phase, it is advisable to consolidate the raw data sheets and confine them to the specific areas of interest through clipping. This approach not only expedites the process but also helps mitigate the risk of tool failure or prolonged processing times, particularly when dealing with extensive study areas. Creating a clipping layer, essentially a polygon delineating the study area, is recommended regardless of the area's size. This clipping layer can serve as a designated processing area for the tools employed in raster calculations or subsequent processes.

Furthermore, gaining access to archaeological and historical information proves indispensable, encompassing data pertaining to site locations, roads, cities, and other anthropic elements pertinent to the study. While much of this data is not readily accessible, it may occasionally be procured through heritage protection agencies or found within monographic works authored by researchers who have compiled archaeological inventories of cities or conducted regional survey studies. The primary challenge lies in referencing the spatial locations and addressing potential coordinate transformations, as previously mentioned.

Additionally, ensuring the reliability of the archaeological data and methodology is paramount. Utilizing aged data often entails dealing with imprecise site locations and multiple sources featuring disparate methodologies.

Consequently, it is imperative to meticulously verify the reliability and accuracy of the data prior to embarking on any analysis. Failure to do so jeopardizes the validity of assumptions underpinning the analysis and modelling. Although this aspect is typically underexplored in publications due to editorial constraints, it warrants due consideration. Generally, the more familiar a space is and the greater the body of research available, the more dependable the estimates will be.

In archaeological studies, it is common to find cases where multiple sites have been duplicated due to different researchers recording them under different names or troponins, despite referring to the same site. For example, one author may use a polygonal grid to identify different archaeological remains within a group, while another author may use points or polygons to represent concentrations, resulting in variations in the number of sites. This discrepancy highlights the importance of carefully reconciling and cross-referencing archaeological data to avoid duplication and ensure accurate analysis.

Once access to a digital model and verified archaeological data is established, the next challenge lies in understanding the ancient mindset. While the specific details of this process are outlined in each applied model, it generally involves critically analysing literary sources and conducting research within the historiographical tradition. This task requires extensive effort to identify, analyse, and compile relevant data. Our interest extends beyond qualitative value; we also strive to incorporate quantitative aspects into our models. Quantitative information, such as the number of workers, wheat production in a particular area, or travel time between locations, can significantly contribute to our models. However, it is crucial to exercise caution when utilizing isolated examples from sources and test their applicability within the broader context of the models.

In cases where only qualitative information is available, GIS can be advantageous. As discussed in the reclassification and raster calculator sections, qualitative data can be converted into quantitative form. By mapping places that meet specific criteria and understanding the potential relationships between these constraints, we can theoretically extrapolate and apply model calculations to different case studies. For example, if multiple cost analyses indicate that certain paths have values twice as high as others do, it is not the absolute value that matters, but rather the proportion. While we may not know the exact travel time for a specific path, it is likely to be approximately twice as long.

It is important to acknowledge that there are other relevant factors that may not be extensively documented in the literature. This could be due to insufficient preservation or study of certain aspects, or it may stem from a particular perspective or paradigm prevalent at the time. The available information often represents the viewpoint of the elite, while the mindset of poorer individuals may be less well-documented. Additionally, there are societal norms and paradigms that were unquestioned during the historical period under examination.

While an in-depth exploration of these issues could fill an entire book, it is crucial to recognize that we will only address some of them to the extent they apply to the models we are investigating. To further complicate matters, we must also consider the criticism regarding the reliability of ancient sources and their inherent biases, as well as the applicability of general principles to the realities of distant territories, both spatially and temporally removed from the originating precepts.

Therefore, the cornerstone of these models lies in historical knowledge and the need for caution in adopting assumptions from sources while being mindful of their limitations. It is through this approach that we can address gaps in our understanding. The quantitative references, which hold significant importance in these models, should be regarded as examples from a specific time, providing more accuracy compared to our present understanding of that particular aspect. However, the models we develop are theoretical approximations of a far more complex reality that cannot be easily measured empirically.

There are two perspectives to consider in this regard. Firstly, the level of precision we aim for in the model. As previously mentioned, it is not about achieving a 90 per cent accuracy, but rather about discarding data that does not align with the intended purpose. Secondly, if GIS analyses were employed to generate historical models without taking historical context into account, the results would be akin to applying contemporary mentality, rendering them incomparable. This highlights the importance of continuous refinement of models through constructive criticism from other scholars.

The fourth group of essential elements for analysis consists of current geographical aspects, which will be further discussed in the subsequent sub-points.

The resources that can be used in these models are diverse. Understanding where they are located and how to work with them is crucial for conducting a proper analysis. We cannot delve into listing all the resources that depend on each specific country or region here. However, in the following sub-sections, we can provide examples of the most important ones and the issues and themes associated with each.

One final note in this regard: the current trend is towards having more and more resources available in open access. This means that we can access them freely and easily. While this is a great advantage, it also comes with its challenges. The main challenge is the management of this information and remembering where it is located or can be found. In other words, when working with various types of cartography, we will often have access to current and highly precise data. However, in historical studies, there may be times when we need data from different times. For example, in a purely archaeological application, we might examine old satellite images to track the changes in a site over time. Access to this information should ideally be free, but sometimes we have accessible resources that we are unaware of. We will now proceed to enumerate and define the most important ones.

4.2 Cartography and Historical Photography

Historical photography and cartography are valuable resources for reconstructing past landscapes. However, access to and resolution of such cartography can vary. It is possible to be aware of the existence of a particular map but lack access or encounter insufficient resolution to effectively work with it. To address these challenges, the next step would involve geo-referencing the historical cartography to match the current geographical data. Geographic Information Systems offer specific tools for this purpose. For detailed guidance on geo-referencing, readers are advised to consult relevant literature (Miraglia 2016; Cascón and Alberich 2021).

Historical photographs are a useful resource within the realm of historical cartography. When working with such information, whether it is hand-drawn maps or ortho-photographs, it can significantly contribute to the reconstruction of past territories. However, caution must be exercised when geo-referencing cartographic maps, as there will always be a degree of error. Perfect geo-referencing of historical cartography is not possible due to limitations in technology, especially when working with maps from earlier periods. When overlaying a historical map onto the current one, it is important to acknowledge the small deviation and understand that exact corrections may not be achievable. Therefore, reference data should be interpreted with care. For example, if a specific place name is provided, it may not correspond precisely to the location indicated on the map but rather to the surrounding area.

In contrast, ortho-photographs present a different scenario, as they are actual aerial photographs of the Earth's surface. They provide valuable information not only in terms of identifying archaeological features but also in their use for remote sensing applications. For instance, in Spain, there are aerial ortho-photographs from 1956 that can be particularly useful due to their pre-heavy machinery era, allowing researchers to identify traces and references that may assist in GIS analysis or understanding past landscapes. Historical cartography can be especially informative for assessing landscape changes by comparing a series of historical photographs or maps, provided they are of sufficient quality. This topic will be revisited in the section on topography modelling.

The significance of historical cartography should not be undermined in general terms. As a valuable resource, contemporary ortho-images or maps prove highly useful, as will become apparent in the following points. Within our field of expertise, we hold a stake in understanding what preceded the present. Notably, we possess the advantage that the majority of agricultural transformations resulting from the introduction of tractors coincide with the era of aerial surveys, making this information invaluable for comprehending landscape changes. Nonetheless, the older historical cartography holds particular significance contingent upon its quality, allowing us to compare spatial configurations, activities, and even site names through preserved toponymical information.

As can be seen, the use of historical cartography is not limited to the geo-referencing of ancient maps, which undoubtedly has its significance. In this regard, access to resources such as historical ortho-photos is highly relevant for a multitude of studies. As a general practice, it is advisable that whenever a study of any nature is undertaken, historical maps with higher resolution and easy accessibility are reviewed. This approach allows us to gain an overall perspective of how changes have occurred. In the specific case of Spain, we often work with the cartographic series produced by the US military in the 1960s, which represents a period prior to significant urban expansion. This provides insight into what the countryside looked like at a previous point in time, offering a clearer imprint of the landscape and a territorial definition somewhat closer to that of the pre-industrial era.

Nevertheless, other resources are of interest, such as satellite images. Specifically, we have access to free images like those from Landsat, which allow for the monitoring of land with very short time intervals. Although their primary use is in line with environmental management, they can also help us understand changes and trends in the territory.

Finally, it is essential to understand that this series of cartographic resources and materials are intended for use from two different perspectives. On the one hand, they serve as contextual data, enabling us to verify information and ensure the validity of previous assumptions. On the other hand, they serve as a foundation for constructing models. We have previously emphasized the importance of working with raster data, especially for generating a digital terrain model. This forms the basis of nearly any model, as it allows us to simulate the Earth's surface. There may be instances where we do not have models with sufficient precision for our study, but there is no need for despair, as there are always options. One of them, although more challenging and time-consuming, is precisely the use of these cartographies and historical data. We will specifically revisit this issue in Chapter 8.1, focusing on the historical restitution of topography.

4.3 Hydrology

Hydrography is a valuable resource when working with models, and it plays a crucial role in reconstructing landscapes and territories. It is not limited to rivers but encompasses all aspects related to water features. Surface water runoff from rainfall and the influence of groundwater flowing into springs and supplying rivers are significant agents of erosion and deposition that shape the landscape (Vivoni and Richards 2005). Therefore, changes in hydrological conditions since Roman times are more likely to be detected and studied. Coastal areas and their transformations are also of interest, as they are highly dynamic due to various factors such as currents, waves, tidal events, and storms (Lufafa et al. 2003). First-hand information about hydrological features is crucial for studying the territory comprehensively. Understanding the main water bodies, distinguishing them from secondary ones, and investigating lakes are all important considerations. It is especially valuable to assess the impacts of anthropic actions and to determine what natural features may have existed in the past.

Streams and smaller watercourses are also significant because they can impede mobility during analysis and may have economic implications, as we will explore later. Additionally, studying watersheds, which represent the total area occupied by a river valley and the flow of its waters, is beneficial. Ancient civilizations recognized these watershed boundaries as territorial barriers (Dilke 1971). While general spatial planning layers often provide accessible information on watersheds, it is possible to generate such data relatively easily if it is not readily available. Hydrographic study processes, as detailed earlier, can be employed to generate this information. Furthermore, there may be situations where specific circumstances necessitate modelling the hydrographic basin of a particular stream, which can be useful for analysing areas of influence or territorial limits where these watersheds may act as geographical barriers (Zheng and Baetz 1999).

When working with hydrography data, it is crucial to ensure that it is correctly geo-referenced and aligned with the appropriate coordinate system. Even a slight displacement or incorrect alignment can have significant consequences for analysis, particularly in the case of rivers and streams for mobility analysis. A few metres of displacement can render the entire model useless, as it would not accurately represent the limitations of movement.

Various hydrography tools are available that enable the generation of river networks, streams, and watersheds based on a digital model. If this information is not readily accessible or if higher-resolution data is desired, these tools become invaluable.

The process typically involves multiple steps but is relatively straightforward. The initial step is to use the fill tool, which fills any depressions or holes in the digital model. This step is essential because subsequent tools calculate the direction and number of cells that contribute to each cell in the raster, simulating the flow of water across the surface. If the tools encounter a low-lying area, they will treat it as a potential lake. However, caution must be exercised in areas with actual lakes to avoid their elimination. Next, the flow and flow direction tools are used to generate a raster where water accumulates. Primary locations represent rivers, while secondary locations represent streams. These values can then be reclassified using appropriate tools to extract useful information layers, as we will explore in detail later. With this information, we can easily determine the path of rivers and streams. By classifying the flow values, we can identify different levels of streams. The raster values indicate the number of cells contributing water to the selected cell, and multiplying this value by the pixel size provides an estimate of the total area discharging water at that point. This is particularly useful for hydrological potential analysis, which we will discuss further.

Additionally, it may be of interest to calculate the catchment areas, which represent the boundaries of the areas discharging water to a specific point. For example, a river's catchment area encompasses all the water that ultimately flows into its mouth. Specific tools are available to calculate catchments, but it can also be done manually by creating a point at the desired location. Each discharge point will generate a watershed, which is highly useful when calculating small catchments such as streams, which are typically not readily available in external data sources (Khatami and Khazaei 2014).

As demonstrated in this concise section, hydrological resources can be accurately generated based on the topography of the terrain when direct access to such resources is unavailable. However, limitations arise in regions with significant historical transformations, such as densely populated areas or when investigating specific resources like groundwater and springs. In these cases, it becomes necessary to consult external sources, seeking expertise from hydrographic specialists who can provide the required information. Consulting maps and studies of hydrographic basins are relevant to our study area and proves beneficial, as this information is typically available to varying extents.

It is crucial to exercise caution when working with this data, particularly when dealing with a small area on a large scale. In vast regions, slight variations in river channels may not significantly affect mobility analyses. Crossing a river ten metres earlier or later may yield negligible differences. However, in specific areas, such changes become highly relevant, necessitating the reconstruction of the site's topography and assessment of potential modifications over time. Historical mapping, as discussed earlier, can serve as a viable solution to addressing these challenges.

The case of water presents a particularly intricate scenario for modelling, given its vital importance as a resource derived from diverse sources. Typically, water was not directly obtained from surface bodies like rivers or streams but rather stored through small dams for various purposes such as household supply, thermal baths, or irrigation. Incorporating these variables into our models proves exceedingly challenging, primarily due to the scarcity of precise data, as mentioned earlier, as well as their variability and the multifaceted uses of these resources. It is essential to approach these challenges with caution, recognizing the limitations posed by data availability, as presuming the existence of comprehensive layers to construct the model can lead to erroneous conclusions.

For example, while terrain slope might be deemed a significant factor in determining the placement of archaeological sites, its relevance may diminish when analysing larger areas, especially considering potential alterations such as quarrying or terracing. However, pinpointing potential wells presents an additional layer of complexity, as archaeological data and detailed studies may be lacking in certain instances.

One of the primary challenges in modelling these influencing factors lies in the nuanced nature of water consumption, which historically required relatively lesser quantities sourced primarily from springs, wells, or constructed cisterns. While some springs may have been preserved and their locations studied, GIS layers often contain only the most prominent ones, omitting many others. Similarly, the identification of potential wells necessitates robust hydrogeological investigations to ascertain the locations of underground water bodies and evaluate accessibility. However, this data may be subject to significant temporal changes, particularly with regard to aquifer depletion and overexploitation leading to declining water levels.

Furthermore, the construction of rainwater collection cisterns introduces yet another non-modellable criterion, as they can be built virtually anywhere and under various conditions. Consequently, integrating these dynamic elements into our studies poses considerable challenges, highlighting the need for meticulous consideration of available data and recognition of the inherent uncertainties in modelling such complex systems.

In light of these considerations, studying water supply and access for human consumption is extremely challenging to model due to the lack of data, its changing nature, and the general fact that scarcity can often be addressed through rainwater collection. Only in the case of predictive models could we assess the greater or lesser viability of a site based on its access to significant quantities of water. Even in those cases, the issue of using aqueducts for this purpose would introduce further complexities.

4.4 Geology and Soil Science

Geological data or edaphology (soil science), can be valuable for land and environmental management. Typically, geological data is available on 1:50,000 or 1:25,000 scale sheets. While the data may be old, the challenge lies in interpreting it correctly. This is mainly due to the fact that the maps are usually made by sheets in which the criteria for each classification do not have to be the same as in the next one. In small areas, it is not a problem, but if we work with large areas and we do not have a map that combines these sheets, the problem of adding this local information increases.

From a historical perspective, understanding the typology of soils and specific geology can be complex, and it may require further knowledge and research to fully comprehend the analysis. However, even though it is not historical information per se, geological data serves as an informative source that helps specify and understand how a particular space and territory function. In certain models, this information becomes crucial as it serves as a criterion for selecting potential areas. While we may not be geologists, we should not be hesitant to read and explore this type of information. We may not immediately recognize a specific soil type, but by conducting research and having knowledge of the study area, we can often establish parallels and connections.

An aspect to consider is the resolution of the geological data we have. It may not accurately represent the features we intend to study. For instance, certain areas we wish to examine for specific soil types may be underrepresented, or the geological categorization may be continuously changing, resulting in mixed terrain classifications without the desired level of detail. This issue can significantly affect the feasibility and accuracy of certain analyses. To incorporate this criterion effectively, it is crucial to have a strong understanding of the field, not only the location of deposits but also the ability to visually identify and associate soil and substrate characteristics with the provided maps. In this way we can check, not the veracity of the data, but whether the resolution and concretization is sufficient for the analysis.

Soil science primarily focuses on understanding different types of ground soil and can be used to create maps indicating their suitability for various agricultural uses, such as crop cultivation, livestock rearing, or forestry. This information is particularly relevant when considering economic and social factors, such as why a particular population concentrates in one area over another, which plays a role in determining potential land uses.

Geology, on the other hand, may be less relevant from an economic and social modelling perspective. It is more challenging to associate specific geological strata with economic and social factors, except when considering the use and anthropic transformation of certain materials, such as various types of stone quarrying. However, geological information becomes valuable when seeking to understand how the landscape has been modified, as certain strata classifications can indicate transformative processes.

In the field, we can readily observe these transformations. For instance, the presence of a rocky outcrop indicates the absence of underlying archaeological levels. Conversely, when encountering dunes, determining whether such formations existed in the past, such as in riverbed or river mouth contexts, becomes more challenging (Steinberg 1996).

In these cases, where we encounter these deposits or transformations, we can infer that the space was different in other historical periods and underwent a process of deposition. Simulating these issues allows us to discern between geological levels that have remained unaltered over time and those that have experienced changes, providing a criterion for identifying landscape transformations (Baker 1989).

Access to these resources largely depends on the country in question. Precise resources are typically administered by national agencies rather than being available at an international level. For example, in the case of Spain (Magna 2022), multiple series of resources are accessible. However, it is essential to exercise caution when using such data. Continuous and non-continuous layers are available, and working with non-continuous layers makes it challenging to establish relationships between data from different areas.

We have already discussed that working with geological and soil type resources can be challenging due to the need to understand their methodology and typology. Additionally, access to such resources can be complex, especially when seeking continuous maps that allow for the analysis of large territories. Finally, the precision of the data itself can be problematic, especially in specific case studies.

However, readers should grasp the potential of incorporating these resources into territorial studies like the ones we are proposing. This potential goes beyond interdisciplinary aspects, which involve combining various environmental, geographical, and natural elements. It also extends to an interpretative key. Incorporating these data allows for a better understanding of the terrain, almost as if one were physically present. Of course, someone familiar with the terrain may argue that knowing the territory is not the same as simulating it. We agree with this, in the sense that anyone conducting this type of analysis should be acquainted with the terrain and have a physical sense of it to formulate realistic historical questions.

In addition, it is also true that there are certain aspects of the terrain that a person may not notice without specific training, such as the geology or the soil type. Therefore, their inclusion significantly enhances the way any analysis is conducted. For example, we might be interested in a particular archaeological site that appears to be out of place in the surrounding territory. Upon examining the stones at the site, we may realize that they are not local but were brought from somewhere nearby. With a quick look at the geology, we might discover that there are no viable areas for extracting these stones except near a distant city. In this example, visualizing the data allows us to correlate two spatial elements. If we can also include this as an additional criterion in our analysis, it becomes even more valuable.

4.5 Current Uses

Lastly, we can incorporate a range of resources that fall under the category of current uses, providing us with essential information about the present state of the territory. These resources include various layers of data such as roads, infrastructure, population centres, and current land use, including agriculture and other activities. This information allows us to compare and contrast it with our models.

However, there are several considerations to keep in mind. Firstly, we must avoid mixing current information with historical contexts when we are uncertain about the conditions in ancient times. For example, in a mobility model that incorporates distance costs and wooded areas, it is crucial to determine whether the forests we have today were also present during Roman times. Given that land use patterns may have undergone significant changes over time, it is necessary to exercise caution. Our concern should revolve around not only whether a particular area was wooded during the Roman era but also the possibility that other areas within the same territory may have been wooded in the past but are not anymore. It is important to recognize that the level of transformation in the present surpasses that of Roman times, and the displacement of population centres can result in a more limited capacity to modify the environment compared to the present (Hermon 2006, 2009, 2010).

Another example pertains to the historical network of cattle tracks or roads, which can serve as a reference point for studying mobility since they likely followed natural paths. However, it is crucial not to rely on unverified data, as it would inherently limit the accuracy and reliability of our model.

In general, my opinion is that the incorporation of current data within GIS analysis needs to be carefully considered and justified, taking into account the potential for significant changes over time. Current data can be valuable when it provides information that has remained relatively stable, such as soil or geology. However, even in such cases, we utilize this data to study change. Therefore, while it is acceptable to incorporate gullies to analyse landscape deconstruction, caution should be exercised when using them as a primary conditioning factor in a model.

In the same way, it is often very useful information not so much in the realization of the model, but in the discussion of the results. In general, when making this type of theoretical approximation, we may find that there is nothing to compare the result. The ideal is to use historical information from other times or the same chronology but in similar case studies. For example, we can make a model that gives rise to the study of the best vineyard areas for the Roman period and then compare it with current areas, to see if it coincides. The result will probably be different, due to modern conditioning factors that we can discuss and specify, in this case, the displacement of population centres or perhaps a change in mindset when choosing agricultural areas.

Within this information, we have to take into account other sources of information such as environmental information. So far, we have seen cases where we have information on the type of soil or surface, but in various models, we may be interested in knowing the climatology and changes in the environment. Of course, we have to be careful to know if these conditions were like this in Roman times. Access to these resources can be varied, depending somewhat on the scale of analysis we want to perform. For example, in cases where we are working with very large territories, such as between countries, to look at weather conditions, we can resort to extra-national data, which may be accessible as in the European agency through ECMWF (2022). However, for particular cases, such as knowing the rainfall of an entire region, we generally work with generalist data from other authors. It is common to know how much it rains in a place or the temperatures in winter and summer. These data, as we consult them nowadays to see the weather, come from weather stations.

This information can generally be consulted free of charge, so it is a relevant source of information, especially when we can access historical data. In other words, seeing what the weather is like in a particular area today is not very relevant, but if we can take averages over the last 50 years then the data becomes more valid for analysis. As well as this, the current use of certain resources can be useful, with a bit of luck, by informing us of interesting details such as cartography related to the use of springs, the exploitation of aquifers or current or historical quarries.

Finally, we can find certain data produced by other researchers that may be useful for our studies. Although in our area we are not yet very prone to open data sharing, this is the case in other sciences, especially in the field of land management. In this case, we can find very interesting data for our studies, such as analyses of the suitability of a certain crop. A good example of such open data can be found in Epic (2022). These resources provide us with maps of great interest, such as the greater viability of a piece of land to cultivate a crop, based on current parameters or even a certain type of forest. This information, generated by experts in the field, is very useful because it can help us in imagining what the old landscape would be like (Mollot and Bilby 2008). If we currently have a cereal field, these maps are not going to tell us what type of tree would have grown here in Roman times but could indicate which ones would grow naturally and which ones would not, narrowing down the possibilities.

Current information must be handled with great care, and there are three precautions to consider when using it.

Firstly, it is essential to verify the source of the data in question. Sometimes, we may encounter pre-processed data created for a specific purpose that may not be suitable for our needs. For example, a map of springs and natural water sources could be associated with tourist-related data, highlighting specific locations or following other criteria, such as displaying only potable water sources.

Secondly, one should understand the data's range and precision. The accuracy of data is often determined by its intended use, and as a result, it may not be relevant to our research. An example of this can be seen in transportation networks, where secondary roads or local paths may not be included but could be of significance. The same applies to land parcel boundaries, where lines dividing parcels may exist for administrative purposes but do not correspond to physical divisions.

Thirdly, it is crucial to exercise caution regarding how these materials are structured and publicly displayed. There may be sensitive data or information that is not directly transferable, such as the presence of current infrastructure. For instance, a map of water infrastructure, such as bridges, may lack chronological information. In this regard, it is essential not to mix current efforts with historical ones. If such chronological data is absent, it must be specifically researched and added.

Therefore, when using current data for comparisons with historical data, it is preferable to preselect and process the data ourselves rather than simply analysing it with the default layers provided. The same applies to digital elevation models or surface data. While these serve as the foundation for topographical studies, it is imperative to have certainty about the data's source and the processes it has undergone.

When we refer to current uses, we are talking about cartography and elements of the current territory that primarily serve us for comparison with other elements of the territory. For example, if we want to study how a river has changed over time, we need to start with an understanding of its current state. There are numerous resources available for this purpose, primarily for comparison. Continuing with the previous example, there are environmental studies conducted on rivers and streams, generally to understand river basins for water management with dams. This type of general information can serve primarily in three major ways.

The first group of data would be those used to build our models, i.e., foundational data for constructing our analyses. Anything other than historical or antique cartography would fall into this large group, including the aforementioned points discussed. However, we may have other elements, such as current land parcel data or specific elements we want to deconstruct. The clearest example could be the study of cadastral parcels, allowing us to use the current layer to trace back and understand that some of these parcels are ancient, especially if we have toponymical elements that help us understand this problem.

Secondly, we have the current elements that can be used to construct parameters for our models. We have already mentioned that the fundamental process involves taking a historical question based on the ancient mindset and relating it to real territory data through measurable geographic parameters. In this context, we can have parameters derived from ancient sources and others that are not. This is why we often use current data for our models. The simplest example would be human mobility. A person today and in the past would move around on foot, so taking data on how this is currently done can be extrapolated to the past. Sometimes the data may not be particularly reliable because we do not know the changes over time, but we only have current data. This is the case, for example, with precipitation. We can model a scenario where rainfall is relevant and then use current data on how it is now. Obviously, it could have been different in the past, so we have to discuss it. However, we need to collect current data on rainfall values per area.

Thirdly, we may want the data simply for comparison. In this case, it would be outside the model, meaning it does not form part of it but rather contributes to critique or debate. For example, using data on current bridges or reservoirs as examples of current water uses and potential, comparing them with the past. In general terms, we cannot do anything in the past that has not already been done, which is why this data helps us understand logical questions.

4.6 Method for Correlating Ancient Sources and Current Geography

After reviewing the primary sources of geographical information and their accessibility, the following chapters will delve into the methodology and detailed historical information for each of the models. These models are organized into several options based on the specific historical questions of interest, further subdividing each option according to the main variations identified.

In general, all the presented models follow a common structure. As previously mentioned, our goal is not to showcase the best models, but rather the simplest ones that can be easily replicated and modified. With this in mind, we structure the method into four key points: problems, determining factors, methodology, and case studies.

In other words, irrespective of the chosen model, the most crucial and initial consideration is our historical question. Hence, each chapter begins with one or more of these questions. These questions should be broad and not necessarily tied to specific cases. This approach allows us to focus on general issues rather than specific ones, making the analysis more reproducible.

Each question is accompanied by a historical problem, primarily pertaining to the prevailing perspective at the time. For instance, when exploring issues of mobility, it becomes relevant to understand the mindset surrounding transportation, particularly the differences in perspectives between the past and the present (Berglund 2001). This helps establish the fundamental criteria for our analysis, which are further explained in the following point.

The conditioning factors for each model are mainly related to three aspects: the typology of the subject under study, external factors influencing it, and society's perspective. For example, if we examine a vehicle, we consider factors such as speed, load capacity, and so on. If we focus on visibility, we consider factors like height and defensiveness. Similarly, when studying a particular crop, we analyse its cultivation methods and specific needs. These internal conditioning factors are usually tied to the prevailing mindset, as well as the physical characteristics of the object or location being studied. To describe these conditioning factors, we draw upon information from ancient sources and current geography. For instance, if we know that a certain crop is grown on a slope, we consider it as a significant factor. Then, we translate this qualitative value into a quantitative one by utilizing a terrain slope layer and assigning values that correspond to the level of inclination.

Hence, it is crucial at second place to identify the criteria we will analyse, understand what we can learn about them from ancient sources, and then determine if we can establish a connection to current geographical data. This step is pivotal and, alongside the historical question, helps break down a complex problem into simpler questions that can be individually analysed.

For example, in the context of land mobility, there are primary criteria that govern how individuals, animals, or objects are transported. These internal criteria encompass factors such as thrust force, ground friction, resistance, speed maintenance, and load capacity. On the other hand, external criteria pertain to how these internal factors interact with specific conditions or changes in the environment, such as slopes, river crossings, or marshy areas.

When analysing the susceptibility of archaeological sites or determining suitable locations for agricultural and livestock activities, the criteria are linked to the Roman mindset regarding crop cultivation. These criteria consider the optimal placement of sites based on current geographical conditions. These examples illustrate the process of breaking down complex issues into simpler criteria. Those criteria that have a spatial dimension and are identifiable can be modelled, while those that lack a spatial dimension must still be taken into account to the best of our abilities as constraints that may modify the model once it has been established. This approach enables us to classify negative results based on these predetermined conditions.

Then, in a third section, we elaborate on the methodology of the model. In other words, based on the given criteria, we combine multiple factors in a way that allows us to visually represent issues that may not be easily grasped otherwise. While we can gather data on terrain height, velocity, visual prominence, and so on, comprehending their territorial implications in vast areas with numerous sites can be challenging. Hence, in this section, our aim is to combine these factors to the greatest extent possible.

In general, the models rely on a relatively straightforward technical foundation. The primary focus is on obtaining the data relevant to our analysis and reclassifying it in a manner that converts qualitative values into quantitative ones. For instance, if the average slope falls within the range of 5–10 per cent, we assign it a value of 1, indicating its suitability for vineyards. This transformation is often relative, necessitating an approach that begins with the historical perspective rather than the current one. These ranges have to be applied manually based on historical knowledge. The method usually consists of translating qualitative criteria into quantitative ones. For example, assign values 1 if the criterion is met and 0 if it is not. This is easily done with data reclassification. Then the values are summed with a raster calculator so that where the maximum value is given, it implies that all criteria are met.

Subsequently, in most models, we merge these values from various conditioning factors to create one or more maps that provide a combined representation of the criteria. Although the process becomes somewhat more intricate when dealing with mobility and visibility tools, the fundamental procedures for establishing the analytical foundation remain the same.

Lastly, we briefly present real case studies where these methods have been applied, either by the author or by external parties, along with an updated bibliography. The purpose of this section is twofold: to demonstrate practical applications, which are often easier to comprehend than purely theoretical ones, and to illustrate that the implementation of these specific approaches is achievable without difficulty.

In this book, our primary focus is to present the model itself rather than the specific results it generates. Therefore, we do not find it necessary to delve into the development of individual cases, as it would involve reproducing numerous maps and cartography that are already available in previous publications. However, we recognize the importance of including a brief section to illustrate the application of the model in concrete cases, allowing readers to grasp its practical implementation.

Throughout the book, our approach is centred on the Roman world as a cultural, geographical, and chronological framework. Many of the assumptions and criteria discussed can be applied to various locations. However, given the inherent variability and potential regional differences, the models need to be adapted accordingly. As a result, we present the criteria and methodology in a general manner, with specific issues and applications highlighted in regional contexts.

In the case study section, we always provide at least two comparable models, facilitating a comparative analysis and enabling readers to understand the strengths and limitations of each approach. This empowers readers to make informed decisions and consider crosscutting choices when utilizing the models. In the next chapters, we will develop each model using the method already shown.

5

Land Use Models

In the first application chapter, we analyse models that have to do with agriculture, livestock and other land exploitation in Roman times. Specifically, we subdivide the question into three main points. The first is about how to determine the potential agricultural and livestock uses of the land so that we can simulate their uses. The second is on how to relate them to specific archaeological sites, using a specific model. Thirdly, and based on the data from the first two, a historical analysis of the economy and population can be carried out theoretically. We develop the method for each one in a specific way, given that it requires different conditioning factors and specific methodological processes. These models are useful for studying ancient economic and social changes over time, but also to understand the sustainability of landscapes. This kind of method could be new for history and archaeology but is well established for environmental planning (Store and Jokimaki 2003). These models serve as valuable tools for integrating agronomic information and the archaeological record within the context of the rural environment.

If a Roman landowner had the option to choose a specific area for cultivation, the preference for crops would depend on various factors. Some crops may hold greater importance than others, and their locations would be determined based on several considerations. In this first model, we aim to address the following questions: What are the cultivation preferences of Romans? Are certain crops more significant than others? Where would these crops be located? What are the primary types of land use, and can we simulate them in a GIS?

5.1 Optimal Areas for Agricultural and Livestock Uses

To address these questions, we can calculate the optimal locations for growing specific crops or engaging in particular economic activities such as livestock farming. In this model, our focus will be on identifying the historical locations of crops. To achieve this, we need to assess the suitability of different areas for producing specific types of crops or similar activities. However, it is important to note that this approach can be applied to other types of land uses as well. The objective is to determine the most appropriate use of land from a Roman perspective, allowing us to classify it based on potential areas. The underlying premise is that if a farmer engages in polyculture, they will prioritize certain regions over others.

For instance, if a Roman landowner possesses extensive property, they might choose to specialize based on certain predetermined conditions. They would not simply arrive and cultivate any crop but would carefully study the land and select a specific type of crop. Agronomic sources provide us with valuable insights into this decision-making process, offering guidelines on selecting the right land. Notable bibliographical references that analyse the context and perspectives of these agronomic authors include works by Martín (1971), Kolendo (1980), and García (1995).

However, it is important to acknowledge that relying solely on agronomic literature, as a basis for modelling spatial relationships may not be entirely accurate. This is because such literature was specialized and not accessible to the general population during that time. Knowledge transfer was relatively slow, and not all members of society had access to these books. The use of these precepts in a GIS model may be challenging as it is only representative of the elite and not of the entire society.

For instance, in the writings of Columella, we find significant criticism of his contemporaries, as he believed there was a lack of knowledge in the field. He highlights the carelessness in choosing and cultivating vineyards, which he criticizes as being unproductive (Columella, *Rust.* 3.3.4–5). Similarly, there is a widespread theory about the land's sterility, suggesting that once cultivated, it will be lost over time. Columella challenges this notion by emphasizing the importance of fertilizing, ploughing, and allowing the land to rest (Columella, *Rust.* 2.1.5–6). These examples illustrate the limitations of relying solely on agronomic literature and the need for a broader perspective in modelling spatial relationships in this context.

Indeed, the precepts outlined by agronomists were often directed towards the elite class, who had significant resources and may not have been concerned with maximizing yields. It is reasonable to assume that the powerful landowners and farmers of the Roman period constituted a distinct group from the general population. Additionally, the management of agricultural estates was commonly entrusted to individuals such as the *vilicus*, who likely possessed a deeper understanding of these matters (Columella, *Rust*. 1.8.4–8).

When examining agricultural production models within the context of the Roman Empire, it becomes apparent that adherence to strictly Roman practices varied across regions, especially during periods of conquest and Romanization, until a certain level of homogeneity was achieved. Divergences in land use, particularly in agriculture, are well-documented in various provinces influenced by Phoenician, Punic, or other indigenous traditions. Scholars such as Gómez (1996), Domínguez Petit (2004), and López (2012) have highlighted these variations. Similarly, within specific regions like Britannia (Rees 1979) or *Baetica* in Hispania (Sáez 1978), distinct agricultural traditions and practices emerged.

Some of these variations can be attributed to climatic adaptations, as evidenced by Gaul, which exhibits notable differences in production methods, particularly in winemaking. Scholars such as Tchernia (1986) and Brun (2003) have elucidated these regional disparities. Additionally, differences in processing techniques have been observed, such as the use of barrels instead of *dolia* in certain areas, as noted by Marlière (2001).

Hence, it would be erroneous to apply agronomic precepts universally without considering the local traditions and knowledge transmission that occurred in specific regions. It is imperative to recognize that agronomic development encompasses a spectrum of scientific practices that evolve over time. Furthermore, it is crucial to understand that the objective of the model is not to replicate reality precisely but rather to construct a plausible representation based on the perspectives of Roman agronomists. If this perspective aligns with the characteristics of a particular settlement, then we can infer a typically Roman pattern of production and land use, as suggested by Sanchez-Simón (2021). However, if this is not the case, we must refer to other data sources to establish criteria for understanding the diverse parameters at play.

In the Roman world, the decision of a farmer to choose a specific location for agricultural activities would have been even more critical than it is today, particularly considering the limited technical capacity to modify the land.

Although there were instances of land modification (Columella, *Rust.* 2.15.4), it is reasonable to assume that not all individuals followed such practices. Systematically altering the typology of the land, as we might do today, would not have been as widespread during the Roman era (Olson 1943).

Nevertheless, we can relate the information we have from agronomy to current geographical conditions, especially regarding spatial preferences. We will focus mainly on Cato, Varro and especially Columella on choosing uses for certain land, precisely because of the ease or not in using these optimal areas. The process consists of looking at the constraints that agronomists refer to and associating them with crop typologies or other land uses.

5.1.1 Conditions

To select an optimal location for a specific agricultural activity, we need access to information that allows us to compare past and present agricultural practices. Factors such as access to resources like water, the slope of the land, orientation, and sunlight exposure all play a role in determining the suitability of a particular area for specific crops.

To generate a map of agricultural land uses, we consider four criteria, as presented in Table 1. Each criterion corresponds to a spatial variable that can be measured today and reflects the Roman mindset based on the agricultural knowledge passed down to us. Here is an overview of the criteria (Figure 2):

- Slope: This criterion is of utmost importance and serves as the basis for many subsequent processes. A slope map is created using a digital elevation model. The process involves converting the Z-values (terrain height) of each cell into slope values. This transformation follows an algorithm that relates the height of one cell to its neighbouring cells, generating a new value. The slope helps determine the preferred use of plains for cereal cultivation, hills for vineyards, and other uses for different slopes, as Varro said, 'because of these three types of relief there are some differences in what is planted, because it is thought that the best grains are from the plain, vines from the hills, forests from the mountains' (Varro, *R.R.* 1.6.5.).
- Orientation: This criterion focuses on classifying preferred hillside locations based on potential sunlight exposure, which is better in areas

facing south rather than north (Columella, *Rust.* 4.22). Wind patterns also need to be considered, such as strong winds from the east and south in specific regions:

> Democritus and Mago commending the northern quarter of the heavens, because they think that vineyards exposed to it become the most productive, even though they may be surpassed in the quality of their wine. To us it has seemed best to direct in general that vineyards have, in cold regions, a southern exposure, and that in warm ones they face the east; on condition, however, that they are not harassed by the south and south-east winds, as are the maritime coasts of Baetica. If, however, your tracts are subject to the aforementioned winds, it will be better to entrust them to the north or the west wind; but in hot provinces, such as Egypt and Numidia, they will be exposed more properly to the north alone.
>
> Columella, *Rust.* 3.12.6

This preference for optimum crops, like vineyards, to the west and other uses to the north, is determined using a terrain aspect tool. This tool indicates the cardinal direction each cell is facing, helping us understand the orientation of different terrain elevations. Additionally, potential insolation or solar radiation data can be utilized to identify areas with higher sun exposure and those that are shadier. This information is related to the orientation aspect and can be analysed alongside it.

- Edaphology: This criterion focuses on the soil classification according to the FAO classification system. It is known that certain soil types are more suitable for specific vineyards in the study area, based on ancient references. For example in the next fragment, we can see the preference for determining the type of soil and the improvement and care related to vineyards:

> If there is no manure of any kind, it is very convenient to do what I remember from my paternal uncle, Marcus Columella, a very skilled and hard-working farmer, what he did on many occasions: he used to put clay on sandy soils and sand on the loamy and excessively dense ones. In this way, he not only obtained abundant crops, but also very beautiful vines. He said that manure should not be put on the vines, because it corrupts the taste of the wine, and he thought that the best material for an abundant

> harvest was soil piled up at random or from thistles, or finally any other material brought from elsewhere.
>
> Columella, *Rust.* 2.15.4

Soils can be grouped into categories, and their suitability for agricultural use varies. For example, high mountain soils may be suitable for forests but not for agriculture, while salt marshes or saline zones are generally suitable for livestock farming. Arable land comprises soils with varying fertility levels, with more fertile soils typically used for preferential activities like vineyards. However, local peculiarities must be taken into account. For instance, soils that dry out and crack during summer may not be ideal for vineyards but could be suitable for cereal crops. These criteria allow us to analyse the landscape and identify the optimal locations for various agricultural activities based on Roman perspectives and practices. It is important to consider the specific context and characteristics of each region when applying these criteria, as agricultural practices may vary.

- Finally, the proximity of streams and paleo-coastal areas is considered with a catchment area of 50 metres, which is reclassified and converted into a raster. The proximity of these areas allows the collection of wood and cordage, which is necessary and a priority according to the agronomists themselves (Varro, *R.R.* 1.7.9). These materials generally require humid conditions, either salt or freshwater, so their use would be reserved for these purposes (Cañizar 2016). As an example, we can see the use of those materials for vineyard production. The usual recommendation is to produce it on the same farm:

> Now that I have set out the precepts which I thought would be useful for planting and cultivating vines, I must now show you how to find stakes, poles and wicker. These, in fact, are prepared beforehand, as if they were endowments for the vines. Without them, the farmer has no basis for planting vineyards, as he has to look outside his farm for everything he needs. And not only does the purchase price increase the farmer's expenses, as Atticus says, but it is also very difficult to acquire them, as they have to be transported in the winter, which is a very unfavourable time to do so. For this reason, it is necessary to have planted willow groves, reedbeds and regular forests, or to have chestnut trees planted specifically for this purpose. Atticus thinks that each *iugera* of wicker can be enough to tie up twenty-five

yugs of vines; each *iugera* of reed beds is enough for twenty yugs of vines; and one *iugera* of chestnut trees is enough to stake as many of reed beds will yield.

Columella, *Rust.* 4.30.1

The association between these ancient sources and current geography has been addressed previously (Sáez 1988) and verified more recently (Trapero 2021c), so we refer to these publications for further information. In addition to these criteria, many others can be taken into account. For example, the proximity to a city or a major communication route. We have omitted this question, which will, however, be used in later models. The remaining criteria mainly pertain to mobility analysis, involving proximity to or distance from specific spatial elements. While these criteria are significant, their relevance depends on the type of terrain and factors such as population density and proximity to communication routes. Given that this topic aligns with the subject of mobility, it would entail duplicating a considerable portion of the principles discussed in the subsequent chapter on land, river, and maritime mobility models.

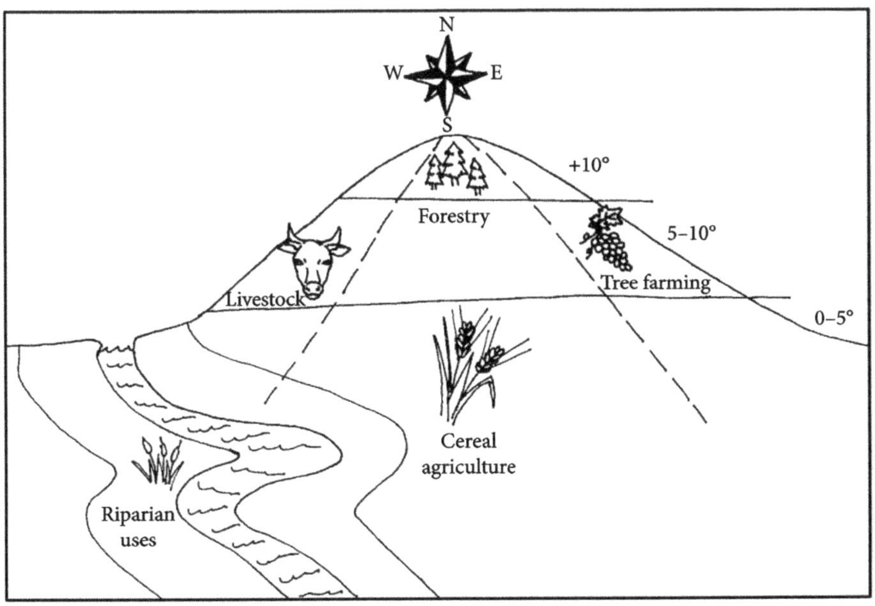

Figure 2 Criteria applied to determine land uses (source: E. Toscano).

5.1.2 Method

Having grasped the main criteria, let us now explore the types of realities they can represent. If we were to model a specific estate to understand its potential uses, we would follow a similar approach to what we have discussed thus far. To create a viability map for crop types, the soil component is typically analysed, along with other parameters such as water availability and climatic conditions. This allows us to determine the suitability of different areas for specific crops. In the ancient world, certain agricultural activities, such as growing barley or wheat, are quite similar, with minimal variation in terms of optimal locations, barring some adjustments in the agricultural calendar.

However, we can identify at least five major groups for land use classification. These groups are based on the Roman perspective and practices.

- Cereal-growing land: This group includes areas primarily used for cultivating grains like barley and wheat. These lands would typically be located on flat terrain with fertile soils.
- Trees: The reference crop for this group would be the vineyard, as it holds significant importance. Optimal locations for vineyards would have good soil quality, generally avoiding both flat plains and steep slopes. Ideally, these areas would also be protected from strong winds.
- Livestock farming: This group comprises areas where cultivation is not optimal but suitable for grazing and other livestock-related activities. Pastures may be shared with tree crops or other land uses.
- Wooded areas for hunting: These areas would not be suitable for cultivation or livestock farming but would serve as hunting grounds. They would typically have characteristics that make them less desirable for other purposes.
- Riparian uses: This group pertains to areas near freshwater or saltwater bodies, such as springs, streams, rivers, lakes, marshes, or coastlines. In antiquity, these areas were cultivated to obtain building materials, basketry materials, and other similar resources. If a space includes such water features, it is possible that nearby areas would be reserved for these specific uses. It is worth noting that when Varro, following Cato, lists the preferred crops, the first is the vineyard, followed by the irrigated orchard. However, the third preference is not the olive tree or cereal fields but the willow grove, which is associated with riverside areas (Var. *R.R.* 1.7.9).

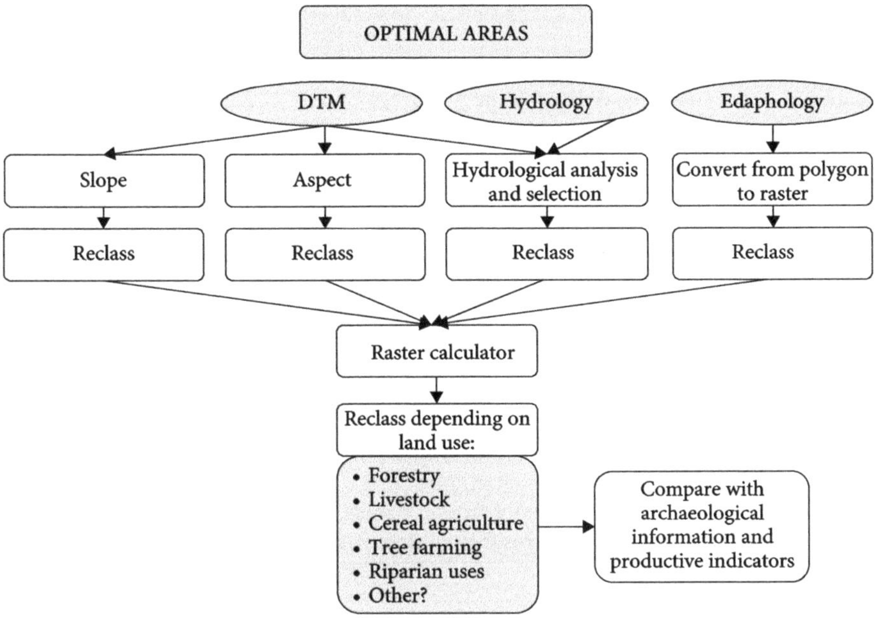

Figure 3 Scheme of land use model. Circles are data layers and boxes processes (source: author).

By categorizing land into these groups, we can better understand the potential uses of a given space from a Roman perspective. It is important to consider proximity to populated areas, as cultivation areas may be concentrated closer to settlements rather than further away. This classification provides insights into the diverse agricultural activities and land uses in the Roman world. This process is explained in the next scheme (Figure 3).

To generate the model, it is necessary to start with a delimitation of the territory to be studied, a DTM and layers of geographic and environmental information, as well as various GIS tools. The methodological process requires a reclassification of the example values in Table 2 so that each criterion has a place in units, tens, hundreds and thousands. The different raster values are combined by addition, so that, for example, the value 1111 corresponds to a flat slope, western orientation, marshy soils and proximity to riparian zones.

With this method, we can easily associate old agronomic criteria with current elements, visualizing each combination and reclassifying it according to the allocation of uses, such as agriculture, livestock or forestry.

Table 2 Example of geographical and environmental elements with reclassification value

Slopes	Between 0 degrees and 5 degrees	1
	Between 5 and 10 degrees	2
	More than 10 degrees	3
Orientation	West and northwest	10
	South and southwest	20
	East and southeast	30
	North and northeast	40
	No orientation	50
Soil types	High mountain soils	100
	Soils with saline substratum, marshes or flooded soils	200
	Soils that are poor because they are acidic or have shallow depths	200
	Soils that are flooded in winter but good for agriculture	300
	Optimal soils for agriculture	400
Hydrology	More than 50 metres from a stream, river or coast	1000
	Less than 50 metres from a stream, river or coast	2000

The method involves categorizing each conditioning factor based on its viability, considering parameters from the Roman perspective. Among these factors, we believe that the type of soil holds particular significance. In certain cases, the soil type can be more limiting than other factors, such as the availability of sunlight, which may have a relatively lower importance. This implies that each type of land use can be classified based on the relative significance of these factors. In this scenario, it would be beneficial to assign a score to each condition, using a scale of 1 to 5, where 1 indicates non-compliance and 5 signifies perfect compliance. Subsequently, the data can be reclassified using appropriate weightings, such that, for instance, soils account for 30 per cent and insolation for 10 per cent of the overall sum. It is important to note that these values need to be reassessed, as we will further discuss in the section on application examples.

We can classify land based on its intended use, designating flat areas for cereal crops, mid-hills for tree crops, and high mountains for activities like livestock rearing or forestry. Location can also be considered alongside other criteria, such as proximity to water bodies, which can be especially relevant for specific purposes.

However, the impact of these factors will depend on the specific characteristics of the terrain. For instance, in the southern region of the Bay of

Cadiz, the direction and strength of the wind can exert a significant influence, potentially damaging crops, as mentioned by Columella (*Rust.* 3.12.6.). Consequently, the weight assigned to each conditioning factor in our analysis will depend on the prioritization we establish. It involves constructing a scale for each factor and calculating their cumulative effect accordingly.

5.1.3 Application Examples

Let us examine how this approach can be applied to a specific case, focusing on viticulture as an example. Viticulture is an economic activity that has received significant attention in the literature, and its spatial conditions for cultivation are well-documented in agronomy. While other examples such as olive trees, livestock, or forestry uses could also be explored, the availability of reliable data for comparing models in those cases may be more limited. However, this does not imply that there are no good examples of ancient landscape modelling based on these uses, which partly form the foundation of the model presented here (Van Leusen 2002; Goodchild 2007; Viitanen 2010).

Our main example is for understanding the best locations for wine production based on the specific agronomic criteria provided by agronomists. This method has been developed for the case of the Lower Guadalquivir region in the southern Iberian Peninsula, known as a potential wine-producing area (Trapero 2016; 2021a). The method takes into account the aforementioned criteria, utilizing relevant information from the native author Columella (Trapero and Lagóstena 2019). Soil types are related to current typologies, and specific conditions such as slope, orientation, sunlight exposure, and proximity to cities, rivers, or specific resources are also considered. The resulting model identifies the best potential areas for Roman vineyards versus areas that are less suitable. In this case, the model is compared with archaeological information and production indicators to assess the consistency of the data. Notably, there can be shifts in productive areas, primarily due to changes in production practices.

Another significant case study is Pompeii, where the unique preservation of the site following the volcanic eruption provides an exceptional opportunity to model these types of questions with reliable data (Vogel et al. 2016).

More recently, the aforementioned method has been tested and validated in publications focusing on the north-eastern Iberian Peninsula, yielding excellent results. Several parallel methods, including the ones mentioned above, were compared, and it was concluded that the most statistically optimal approach is the one that considers agronomic precepts (Stubert et al. 2020).

Similar studies, albeit on a different scale, are being applied to the Gallia region, with a particular focus on viticulture. Although the analysis encompasses entire regions, the methodology employed is similar to the one presented here (Bernigaud et al. 2021).

These examples demonstrate the practical application and effectiveness of the methodology in modelling ancient landscapes and understanding the optimal locations for specific agricultural activities, such as viticulture.

From these examples, one can observe the potential inherent in this type of analysis, although it is currently an area under study. These examples are reproduced in various fields of knowledge and specific themes. We have chosen the topic of wine as our reference, which is, incidentally, one of the most significant. In practice, it can be said that its potential for future use depends on further comparative studies and its application in other areas. Specifically, these examples often address broad areas, as seen in Bernigaud and other authors (2021), or more specific territorial cases. The challenge of bringing these together is an ongoing issue that requires the practice of analysing other researchers' work in our own case studies to validate the processes.

Those examples are from historical periods, but we could add other approaches that use similar concepts for gather and hunter societies, as we could see in the work of Surface-Evans (2012). Those approaches use other parameters such as catchment areas and fauna viability for doing those land use calculations, normally with mobility settings (Nolan and Cook 2012).

5.2 Relationship of Sites to Nearby Uses

In our second model, we aim to establish a connection between the previously discussed criteria and specific archaeological sites, particularly focusing on the Roman period and the *villa* as a representative model for rural occupation and exploitation. However, we often possess information solely about the archaeological site itself, lacking details about the productive area it would have encompassed. In this context, understanding that a building complex is situated on a particular soil or orientation may not provide direct insights. Instead, we are interested in knowing whether the surrounding land near the *villa* is suitable for activities such as vineyards or similar agricultural uses. This raises the question of how we can relate spatial information about land uses to specific places like rural settlements.

To address this challenge, we have developed the following method that associates agricultural uses with the archaeological site, acknowledging that we do not possess precise knowledge of the actual territory of each *villa* as an estate. Instead, we adopt an approach that utilizes an area of influence or buffer surrounding the site, representing an average value of the villa's territory.

In a variant of the aforementioned model, the objective shifts from determining optimal zones to understanding the specific relationship between the site and its surroundings. However, in GIS studies, there is a fundamental difficulty in aligning geographical information about a given territory with the specific case of an archaeological site. We often have separate layers for soil types and archaeological sites, but establishing a meaningful connection between the two can be challenging.

One possible solution is to utilize the value extraction tool, which allows us to extract information from the raster layer representing soil types. Rather than treating it as terrain height data, we extract the specific soil type associated with each site, along with other criteria such as slope or orientation. However, this alone may not be sufficient, particularly when studying land use from an agricultural perspective. While the country house or *villa* may serve as the *pars urbana*, the focus of this model is on understanding the surrounding area. Though we can visualize the general map and locate the points, it remains difficult to assess whether an area has a higher proportion of a particular soil type or conditioning factor compared to another site. Hence, the challenge lies in modelling and correlating the spaces near a settlement.

By generating an area of influence around each *villa* and extracting values within that area, we can calculate the percentage of land for each specific condition. For instance, if we aim to determine the predominant geology types surrounding a site, we can cut the raster using the area of influence and extract the values, allowing us to employ statistical analysis to identify the prevalent geology type in that area. This approach enables us to understand if a certain space has a more specific type of use than another does, or if proximity to clay for pottery making indicates a higher likelihood of it being a pottery centre.

The same methodology can be applied to assess the potential use of other resources such as hydraulics. By comparing various layers of information, we can evaluate which *villa* within the area of influence has a greater proportion of land suitable for riverside exploitation. This analysis provides insights into the characteristics of the surrounding space, the territorial organization, and the significance of various conditions in shaping land use patterns.

5.2.1 Conditions

There are two fundamental parameters to consider in our models: the size of the farm and the human capacity and suitability to move. To address the size of the farm, we can use the shape layer of the city or property limits as a trimming limit. Alternatively, if such a layer is not available, we can create a buffer around a representative area of the case study. To determine the size of the buffer, it is ideal to conduct an initial study of the average distances between villages or other entities. Thiessen polygons can be used to calculate the area occupied by each entity and then average the values to obtain a representative size. Another approach is to use a tool that calculates average distances between sites, such as the cost allocation tool. This tool generates edges based on topography and chosen centres. The important aspect here is to establish a territorial boundary for our entities.

However, this approach may not be logical within the settlement structure, as there might be missing villages or other entities. Therefore, in areas where the existence of such entities is possible, these areas need to be excluded from the study. This process is feasible when we have certainty about well-studied settlements, and it can help identify potential missing villas. Determining the maximum size of the village in this context can be complex, and results that are more accurate can be obtained in areas with better-studied and well-represented rural settlements.

The second parameter to consider is human mobility within the space itself. While a distance cost analysis around the area can be used to assess the actual area of influence based on topography, we have chosen to simplify the process by using buffering. It is important to understand that each space will have an area where it is easier to move, and individuals tend to move within this zone. For example, in modern times, it is feasible for us to drive from one city to another within 30 minutes and cover a distance of 50 kilometres. However, in ancient times, travelling this distance in a day would have been challenging and not a feasible distance for workers to commute. These rings of useful mobility zones can be methodologically interesting to implement in case studies.

From the perspective of historical sources, it is challenging to find examples where the mentality of mobility is explicitly discussed. Mobility, in this context, generally refers to trade-related tasks and the supply of goods that could not be produced within the countryside, such as tools. The logic suggests that individuals would not leave the farm unless necessary for these activities. Fieldwork, in particular, would have to be conducted in close proximity to the area being worked on, as workers would not be able to travel long distances. If we do not have the exact boundary of the farm, or even if we do, we can define a vital area for each entity based on a distance cost ring. The specific parameter to use would depend on the level of knowledge we have about the area. For example, a later example from agronomist Alonso de Herrera recommends not having wine presses located more than 4 kilometres away from the place of collection, as this could affect the condition of the grapes. This value, though somewhat arbitrary, provides a window of 1 to 2 hours of maximum travel time, considering both the outward and return journeys. However, it is important to note that these values are not extensively studied and would require further research for their accurate application in these models (Figure 4).

5.2.2 Method

The method described here involves selecting each token independently to avoid repeating the process. In the second level of analysis, we create a buffer or area of influence. There are different ways to do this. One option is to create a buffered area of influence, which provides a theoretical approximation of a vital zone, representing the useful area close to a specific centre where certain activities would be concentrated. Another variation of this method is to create rings of different distances to associate certain uses with greater or lesser proximity.

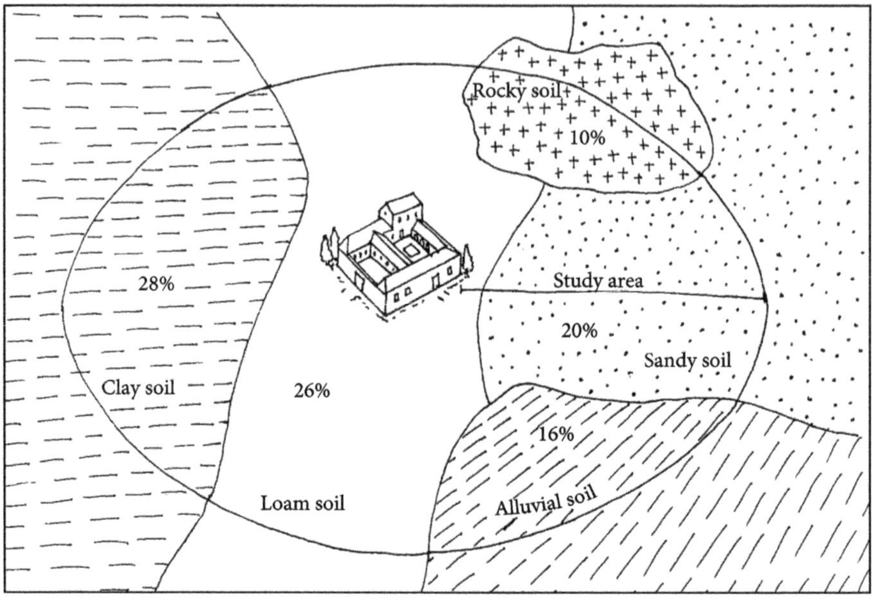

Figure 4 Example of criteria in an ideal *villa* with soil types. The model allows us to know the percentage of each type of soil related to a relative boundary for each productive centre (source: E. Toscano).

Ideally, knowing the boundary of the property or having a proposed boundary would be the best option, which will be discussed in a later model.

In a third scenario, a cost analysis can be used to generate iso-lines that indicate the time it takes to travel from one place to another. This approach may offer more accuracy compared to the first case. However, there is a technical challenge involved. When using a time map generated from the cost analysis, individual adjustments would be required for each area.

The process is as follows (Figure 5): we start by taking the buffer and trimming it to the layer of the features we are interested in, such as specific soil types or land uses. We then perform a dissolve operation to merge all the different polygons into a unified representation.

The buffer tool is straightforward to use. We select an input entity, which can be a point, line, or polygon, and provide a distance to generate an area of influence around it. The result will be a layer of polygons for each entity element. While this tool is commonly used for quick visualization of areas of influence, in our case, we are mainly interested in assigning width to entities that lack it. For example, we can project a point or linear entity onto the surface for further analysis. This tool is useful when we want to transform a line entity, such as a river or a road, into a raster format. Although the line or point entity is converted into a raster, it will not have a surface. This tool can also generate areas of influence or perform small area studies to identify overlaps. Multiple rings can be created, not just at a single distance, but at various distances. This is particularly useful for understanding distances or generating surfaces that can later be converted into rasters. For example, if we want to artificially generate the basin of a river, we can create several areas of influence with increasing elevation values.

Next, we employ the field calculator to append fields and compute the percentage of the total area occupied by each feature. Establishing this accurately can pose complexity, particularly if different distances were utilized for delineating each site. To compute the total area of a circle, the value of the polygon representing that area must be retrieved and incorporated within the table. Subsequently, each percentage can be exported individually, or

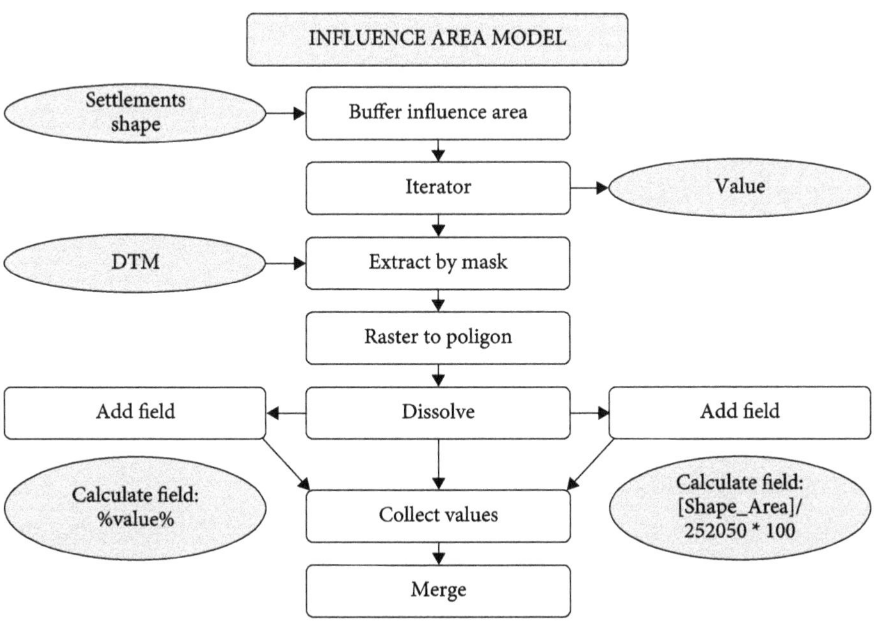

Figure 5 Scheme of influence area model (source: author).

alternatively, the collect value and merge tools can be employed to amalgamate the data. The resultant values can then be exported back to the original layer. Statistical tools such as Excel can be leveraged to generate graphs or conduct further analyses.

As can be seen, this is a more complex technical process compared to the previous cases. However, once automated, it can provide valuable insights into understanding how the territory is organized, especially the relationship between spaces and sites and their immediate surroundings. The summary of the process is outlined below.

5.2.3 Application Examples

This methodology has been developed and applied in specific case studies during my PhD (Trapero 2021a). The distance between the villa, which is approximately the size of a *centuria* (a radius of 283 metres), has been chosen as the zone of influence. However, this value may need to be adjusted based on the density of sites in each study area. The zone of influence method involves using various elements such as a layer of archaeological sites and a raster with criteria generated in the previous phase.

To streamline the process, the Model Builder feature and an iterator are used to interconnect different tools. The iterator selects each area of influence independently based on a value, typically the numbering of the villages. Then, the data from the geographic raster within each area of influence is extracted, resulting in a cut-out of the desired area in polygon format. The polygon is then transformed into a raster, and a dissolution tool is applied to unify the polygon fragments based on their value. Two fields are added, and the field calculator is used to populate these fields. The first field returns the %value%, which corresponds to the original data number for later linking. The second field uses a simple rule of three formula to calculate the percentage of each type of exploitation within the total area of the zone of influence. Finally, the values are compiled and merged into a single layer.

As a result, the original settlement layer is enriched with fields indicating the types of existing farms and the percentage they represent. In this case, the method has been applied to understand the percentage of potential uses of the sites based on soil types, geology, and water resources. However, it should be noted that there were challenges with data quality, particularly in the resolution of the soil type layers, leading to misinterpretations in areas close to marshes. This experimental method requires further refinement and future studies.

While the method is novel, it needs to be compared with other examples and refined. The method could be useful for delimitating the so-called 'empty landscapes' (Campana 2018). Examples such as Goodchild (2010) provide reliable approximations that can serve as a basis. Additionally, there are other relevant examples where the focus is not on analysing soil types directly but rather on understanding the effects of their use (Gilgen et al. 2019), which have the potential to be applied in a similar way to the proposed method. This is a newer application with few comparative examples, which makes it a very interesting field for further study and new models to review and improve upon.

The intention is to apply it to other case studies by altering the radius parameter through which this type of analysis can be conducted. While there may be other examples of the same, the truth is that this analysis has limited comparability as it represents a new line of action. Its potential is significant, as we have applied it to a town to understand its location in relation to nearby resources, enabling the comparison of specific entities with polygonal areas (Rivers et al. 2013). However, its impact could be much greater, as the idea is to extract information of various kinds in different areas and compare them as a whole. In other words, it could be applied to territories in different cities, once they are defined, to quantify the useful areas for each economic activity. The same can be done to generate predictive models, for example, in the case of a Roman road, where we do not know the area that was used. In this regard, an intermediate space could be calculated and an attempt made to correlate it with the main resources and constraints.

5.3 Production Areas and Population

An issue in the study of Ancient History revolves around understanding the potential of space for production and its capacity to sustain a population. The presented model offers a comprehensive approach that enables us to comprehend both the economic implications of an activity and the demographic aspects of rural settlements. Various methods exist in the literature for calculating the population of such settlements (Zorn 1994; Carreras 2014), which can be approached from two perspectives: the urban environment and the rural environment.

The urban method involves determining the area occupied by city walls and calculating average population densities for different areas within the city. Conversely, estimating the minimum population on a given farm considers the number of people required for specific economic activities, establishing a relationship between the economy and the population per unit of work (Carreras 2014).

This historical question arises in numerous contexts, particularly when attempting to understand the population density of rural areas. A key challenge lies in determining the type of entity under consideration. For instance, an average Roman *villa* would accommodate a larger population than a small farm managed by a peasant household. The same applies to larger entities like vici or similar settlements (Scheidel 2008).

Factors such as property size, labour force composition, and cultural attitudes significantly influence the analysis. Weber's work (2008) provides a foundational framework for understanding these relationships, while insightful books by Hollander (2019), Verboven and Laes (2016), and Carlsen (2013) shed light on the associated mentality. However, it is important to consider the specific period, as demographic patterns differ significantly between the High Imperial and Late Imperial eras (Rathbone 1991; Banaji 2002). Estimating the minimum productive area required for a basic rustic farm determines the essential tasks involved (Kenny and Fussell 1996).

The aim is to compare theoretical estimates of productive and economic demographics and apply them within a spatial model. If we have knowledge of the productivity of a given space, average consumption per person, and the required labour force, we can estimate values that suggest potential uses. This approach is particularly valuable when combined with the previous model's understanding of farm boundaries, enabling estimation of the population on the farm and exploration of potential land uses.

Establishing a specific model for the entity type under investigation simplifies the methodological process. Moreover, there are successful models in the literature, although not identical to this study, that offer complementary insights (Lo Cascio 1994; Bintliff and Sbonias 1999; Goodchild 2005; Hin 2013; Verhagen et al. 2014).

We will focus on the example of a Roman *villa* as our case study, as it provides sufficient information from historical sources. These sources allow us to gather knowledge about agronomy and the necessary labour activities associated with the villa. With this information, we can establish a relationship between the size of the land surrounding a farm and its potential uses. By conducting a combined analysis that calculates the hypothetical boundaries of a farm, we can estimate the number of farms. This approach becomes even more valuable when applied to a larger territory, enabling us to understand the population involved in specific economic activities and the relationships between them. Now, let us return to the conditioning factors.

Although our model is purely theoretical, it is also based on the ancient mindset. In Roman agronomy, the necessity of a certain number of people for cultivating a piece of land is well-documented. For instance, on the one hand, cultivating cereal grains requires a workforce with a minimum consumption level. On the other hand, activities like vineyard cultivation, which may be a preferred focus, typically require only one person. Additionally, there are other economic activities related to vine production, such as forestry and the management of production areas, which involve the use of wood materials.

The actual numbers of people required would depend on the total cultivated area and would have a reciprocal relationship. In other words, as the number of people working in the vineyard increases, so do the costs of cultivation to sustain them, as well as the need for a larger area for riparian production to obtain wood and ropes for the vines.

It is important to acknowledge that these conditions raise certain complexities. People would not solely consume wheat or bread; their diet would be more varied. However, this basic estimation provides a foundation for assessing the significance of each activity. Although these calculations are theoretical approximations, numerous studies have shown that quantifications of this nature can yield significant results (Duncan 1974; 1990).

To calculate the fixed population of a farm, we specifically consider the people who would work on the farm throughout the year. However, certain activities, such as grape harvesting, may require hiring additional labour on a temporary basis. Therefore, the number of people in this calculation does not include temporary or migrant workers. It is important to distinguish between the permanent and temporary labour force. In addition to historical factors, we also need to consider current determining factors, such as the typology of the *villa* under analysis, to obtain accurate numbers for our model.

5.3.1 Conditions

First, we need to decide on an area of analysis. The ideal would be to have a well-contrasted space to be able to calculate the average and to be able to apply it, perhaps to a *villa* and then extrapolate the average. If we have a well-defined case study, one way could be to apply a model such as the previous territorial limits.

However, we could also calculate the optimum zones and then, within the area of influence of each site, analyse which places are initially dedicated to viticulture, which to livestock or forestry and, in the model, to calculate the production itself. In this case, we can understand all the economic issues, but also calculate the potential population that would exist in that area and the implications of production, the land area needed for crops, etc.

We can take the basic precepts of agronomists that inform us of types of production as our criteria. The interrelation between these values is what allows us to make the model. We start with the case of viticulture again, since we have a lot of data on the subject. Although we have information such as that from Cato, which implies 13 slaves per 100 *iugera* (*iugera*= 2.518,21 m^2) of wine and 15 slaves per 240 *iugera* of olive trees (El Bouzidi 2003). This is a measure of management, not in itself of production, so it serves as a reference but not so much for making the model (Lo Cascio 2009).

Better to turn to Columella's data, starting with the vineyard. As can be seen below, the concatenation of sources allows us to establish a series of criteria:

- First, we know that for every 7 *iugera* of a vineyard, a full-time foreman is required (Columella, *Rust.* 3.3.8.).
- That the average wheat consumption per person is estimated as approximately 51 *modii* of wheat per year (Cato. *Agr.* 56.; Duncan 1974: 147), which would be 4.25 *iugera* of land devoted to wheat for the consumption of the worker since the yield of Roman wheat is usually

three to one quantities of grain (Columella, *Rust.* 2.9.1.). This value is increased by adding the seed that is necessary for the next harvest and a biennial system of crop rotation (García 1979), which would amount to 12,276 *iugera* for the same production.

- At the same time, we know that to work 200 *iugera* of cereal, eight people and two oxen are necessary (Columella, *Rust.* 2.12.7–9).

With these simple ratios, we can average the number of people necessary for this work. Moreover, in the case of the vineyard, we know that it was necessary to have an area dedicated to cord wood for staking the vines, specifically one yard of wicker is enough to tie 25 yards of vineyard, one yard of reed for twenty yards of the vineyard with chestnut being the same as the reed (Columella, *Rust.* 4.30.1.).

Therefore, if we take the average of one vineyard worker for 7 *iugera* and eight workers for two hundred *iugera* of cereal, to which we add their possible wives and children by doubling this value. Also adding the supervisor and his wife, we can estimate the total population, but also of the total useful area for these activities (Figure 6). Of course, it must be understood that there would be many other tasks necessary, but at least these would be the people necessary

Table 3 Theoretical estimation of hectares and workers needed for each agricultural activity based on the above-mentioned sources

	Hectares	Workers
Per 7 *iugera* (0.251821 ha each *iugera*)	1,76	1
Ha of wheat that can be managed by one worker	6,30	1
Ha of wheat needed to feed one worker (with seed and biennial rest)	3,09	1
For 10 ha of the vineyard as an example	10	5,7 (6)
Wheat to feed the previous vineyard workers	18,54	2,9 (3)
Wheat feed for the wheat workers themselves	18,9	3
Cord wood (1 per 25 vineyard *iugera*)	0,4	1*1
Reed and chestnut wood (1 per 20 vineyard *iugera*)	0,5	0
Additional population (family x ½ workers*2 + 2 *vilicus* and *vilica*) with extra workers and family to feed them	58,71	19
Total ha and population estimated	107,05	32

**1* Cato recommended a person as a guardian of those areas (Cato. *Agr.* 11). **2* We estimate that workers account for half the population as many could be slaves or have children. Also in the calculation, a worker should eat more, so a method to estimate the hectares could be made using this proportion.

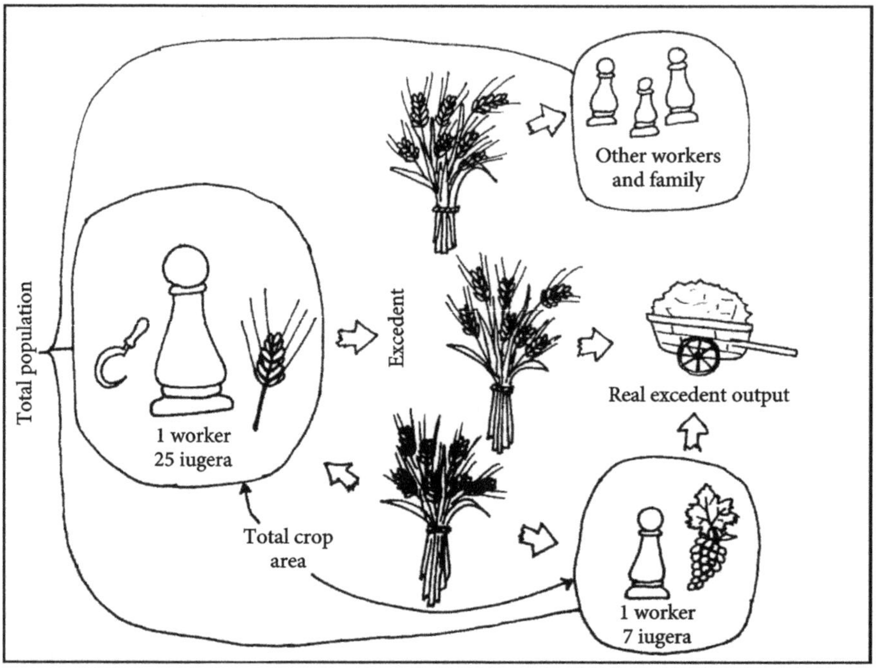

Figure 6 Scheme with the criteria used in the model. The total output is divided into the consummation of the workers and the real output (source: E. Toscano).

for a farm with cereals and vineyards. Other activities, such as olive growing, require proportionally less work and people involved.

5.3.2 Method

The model in this case does not necessarily imply a complex GIS analysis, but rather a statistical analysis. The use of this kind of approach is not new in history and archaeology (Voorrips 1987). From this model, we had to relate each typology of the optimal area that we have analysed by a multiplier of the number of people or economic costs involved. In this way, the model here is simpler than the previous one and does not require the use of additional tools.

It is indeed interesting to recognize that the application of this method relies on establishing a spatial relationship from a specific location. Variations of this method can be implemented in a more generic manner. For instance, if we know that a certain area of agricultural land requires a certain number of people; we can calculate the total area required for various activities, not just for a *villa* but for an entire territory. This enables us to estimate the minimum number of people needed to cultivate those spaces. These averages provide an approximate estimate of the population engaged in agricultural activities within the territory, which can be highly valuable for population studies. While some data are based on my doctoral thesis, primarily focusing on the theoretical approach to using this methodology, we can provide a theoretical example to illustrate the point.

The general method encompasses numerous variables that can be adjusted depending on specific conditions beyond the scope outlined here. For example, the introduction of different crops, specialized labour, or different time frames. Simplifying the approach, the method involves establishing the relationship between three key parameters: people per unit area of production, resource production per unit area, and food consumption by the population. It is generally advisable to start with the area of cereal production on a given farm, as it is relatively consistent. Assuming that bread serves as the basic food staple and that a salaried worker would be paid a similar amount, we have a fundamental basis for our calculations. By incorporating other major activities such as wine production, which requires a greater amount of labour, we can make comparisons with self-sufficient farms (Figure 7).

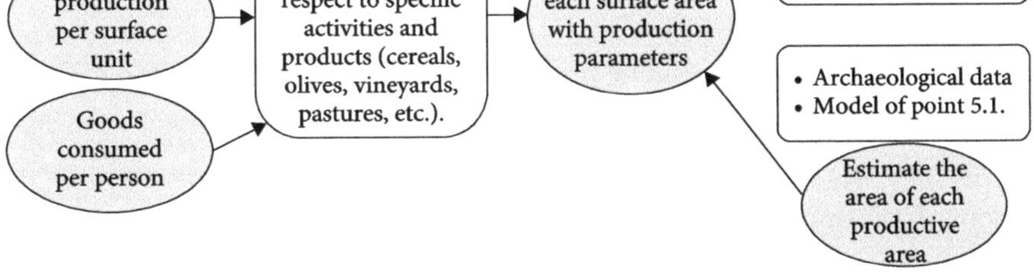

Figure 7 General scheme of how to model the productive area, workers and consumption in agricultural exploitation (source: author).

5.3.3 Application Examples

Similar studies to the one described above have been conducted and can be found in various publications. Generally, these studies have focused on ancient bread production (Amourety 1986) and cereal yield analysis (Comet 1996). They have also been applied in contemporary contexts, such as the organization of work in construction (Courault 2021) or the estimation of seasonal labour in Italy (Erdkamp 2015). A notable example is Rosenstein's demographic study (2004), which estimates the productive needs of a peasant family. Current applications can be observed in case studies conducted in the United Kingdom (Franconi and Green 2019), and variations of this methodology have been employed in multiple case studies, often by the same research team (Martín-Arroyo and Trapero 2015; Martín-Arroyo 2016; Martín-Arroyo and Remesal 2018; Martín-Arroyo and Castro 2019). The approach is predominantly theoretical, necessitating careful contextualization of the data used. However, it provides a more realistic means of addressing issues related to the use of land in agricultural settings. Specifically, these case studies have focused on the study of resource catchment areas such as riparian zones and wetlands. In our particular case (Trapero 2021a), we have attempted to model this approach for ancient viticulture by estimating the areas required for each activity and proposing a minimum potential population.

The results have proven valuable for estimation purposes and as a method to validate field data reliability. In some instances, the calculated exploitation area was excessively large, indicating the possibility of undiscovered deposits. Conversely, there were cases where the estimated territory was insufficient to cover the village's own needs. This miscalculation is of great interest as it suggests that the model does not apply in such cases, either because the activities were not dedicated to those specified or because the site was not a *villa* in the traditional sense.

Currently, we are developing a more comprehensive method for making these estimations. It involves considering not only the production of wine and cereals but also other commodities such as oil, as well as human and animal labour. The concept is simple: determining the minimum productive area required for a space to be self-sufficient (Trapero 2023). This entails estimating the minimum number of people, animals, and land necessary for basic tasks to ensure the village can produce enough to sustain itself. These values, while logical, are intriguing to model as they enable us to assess the realism of a productive model in GIS based on labour requirements and potential land resources. When combined with the method described in Section 5.2, it becomes possible to establish relationships between specific villages based on their productive potential.

Although we have shown partial results applied again to the Roman rural world, its extrapolation is extensive to other areas, such as modelling the conditions of ancient cities or even calculating other aspects. In the case of cities, population studies are generally based on the size of the area, but this is unrealistic when extrapolating the population density of one block to other spaces. Therefore, it would be possible to quantify the area of public spaces, such as the forum or roads, as well as the areas dedicated to luxury houses, which would have a lower population. An analysis of this kind could be more accurate.

Nevertheless, the most significant potential lies in establishing a triple relationship between an individual's consumption, a field's production, and the number of labourers per unit. The rest of the analysis can be scalable depending on the initial data at our disposal and our ability to verify it. Naturally, we have applied this analysis to the Roman mindset, but it is highly likely that many of these principles can be employed in other historical periods and cultural groups. Our perception is that fundamental agricultural tasks, as well as the consumption and productivity of spaces, remain relatively constant. While the land structure and the type of labour may change, other aspects can be fairly consistent. If compared with agricultural sources from the medieval or modern era, this model can easily find applicability in other areas.

6

Restitution of Historical Landscapes

In this chapter, our focus shifts to the reconstruction of the historical landscape, which is a crucial aspect for other models that do not inherently involve complex historical modelling like the previous ones. It primarily focuses on restoring conditioning factors related to topography and its historical transformations.

We acknowledge that the issue at hand is complex, intersecting with other disciplines such as geomorphology. However, we believe that it can be effectively utilized as a historical model using GIS. In the case of Roman times, the changes that have occurred in the landscape are often attributed to anthropogenic rather than solely natural factors (Howey 2007). While it is undoubtedly a combination of both, a deeper study of the natural factors enables us to better understand phenomena such as coastline modifications, estuary enclosures through coastal arrows, or dune erosion and deposition.

However, there is a notable gap in the literature regarding studies on anthropogenic changes. While we have general knowledge about certain aspects, such as increased deforestation during Roman times leading to heightened erosion, the specific details regarding these changes in a given territory, whether large or small, remain uncertain. To address these questions, it is necessary to carefully consider the criteria by which we recognize and measure these changes, whether they are natural or anthropogenic. It is important to note that much of the available information is qualitative rather than quantitative, particularly when comparing historical data. As a result, we can only approximate what the recent past was like before the mechanization of the countryside. In areas with limited demographic growth and minimal natural alterations, the territory may bear a greater resemblance to the conditions of Roman times. We will examine two distinct yet complementary cases in this regard.

The first case involves the restitution of topography, particularly when there has been a significant alteration in the land's surface due to factors such as stone quarrying or natural changes. This entails the reconstruction of the original topographic features to better comprehend the historical context.

The second case pertains to a more historical nature, where we explore the modelling of the potential navigability of a space. This parameter is crucial for understanding communication networks in ancient times, but its assessment can be challenging due to the limited availability of data on both natural and human-induced changes that have affected waterways.

When considering historical topography, we must acknowledge that the methods employed can vary significantly based on the scale of analysis. Large-scale analyses may use digital terrain models to understand mobility within a territory, taking into account major geographic features and variations, such as river positions or hills. Conversely, smaller-scale analyses may focus on specific archaeological sites or rural production elements. The approach used depends on the research objectives, and in some cases, it may involve physical excavation or drilling to determine soil and archaeological strata, which can be costly and time-consuming.

If we are unable to fully reconstruct a distant past, we may attempt to analyse a prior period, such as the modern era before the widespread use of agricultural machinery and modern infrastructure. During this time, the landscape may have more similarities to ancient or pre-medieval landscapes.

It is not so much natural environmental changes that shape the landscape, but rather the actions of humans. The reconstruction of a small-scale space is feasible, provided we have recent documentary information. However, the feasibility varies greatly depending on the location and the landscape under study. In mountainous regions with a rocky substrate, characterized by forested landscapes, changes are minimal, primarily attributed to localized erosion, sedimentation processes, and minor alterations.

In such cases, reconstructing the landscape and its changes is relatively straightforward. In contrast, areas near the coast, especially those affected by intermittent watercourses like seasonal rivers, may experience significant transformations. For instance, these rivers may remain dry for most of the year but can suddenly swell with water during certain periods.

Also, it becomes particularly interesting to study topographical and environmental changes. Nevertheless, the reliability of the method can be challenging, depending on the scope of the study. Therefore, it is crucial to assess the study area comprehensively, considering factors such as the area's extent, the acceptable margin of error, land typology, and terrain characteristics.

In some cases, there are stark differences, such as coastal regions, where substantial topographical changes may result from erosive processes caused by deforestation and other human activities. In these instances, landscape alterations are considerable and may be challenging to reconstruct accurately, especially if the changes occurred in ancient or medieval times and historical records are lacking. In such complex cases, research methods may grapple with numerous uncertainties.

One such uncertainty is the desalination or drying of wetlands, which significantly transforms the landscape. The analysis of transportation costs is also influenced by these landscape changes.

By examining these cases, we aim to shed light on the importance of reconstructing and understanding the historical landscape, which provides valuable insights into past environments and the dynamics of human interactions with the natural world.

6.1 Historic Topography

The reader may find the historical perspective valuable in understanding how the topography has evolved. However, there is no single, straightforward model to address this issue, as it is a recurring challenge encountered in numerous case studies (Hall 2008). While the Earth's surface is subject to constant change, the alterations during the Holocene period are relatively minor compared to larger tectonic and geological movements. From Roman times to the present, with a few exceptions, significant topographical changes are not prevalent (Zevenbergen and Thorne 1987). The extent of change depends on the scale of the study. For instance, when modelling Roman communications in a provincial area, geomorphological changes such as sedimentation or erosion are typically negligible. However, when examining the specific topography of an archaeological site affected by open-cast quarrying, the situation is different.

In the context of topographical changes, two key types can be considered: point events and long-term events. Point events, such as earthquakes or storms, rapidly and significantly modify the terrain. While they can be studied in recent periods, investigating them in the past becomes more challenging unless specific information is available. Conversely, long-term events, which encompass natural and anthropic factors like sedimentation and erosion, can be effectively modelled. Factors like deforestation may exacerbate these phenomena. It is crucial to approach these changes from the present to the past, attempting to trace back to the furthest point in history. Fortunately, the most substantial anthropogenic alterations have occurred more recently with the advent of mechanized agriculture. Thus, utilizing aerial photography and other topographic data enables the restoration of a model that approximates the historical landscape.

Historical information on GIS, ancient sources documenting changes, or medieval and modern records reporting modifications can help identify significant alterations within the topography.

In general, the topography of a region experiences minimal modifications over time. Therefore, the key aspect lies in understanding which areas are more susceptible to alteration, enabling the establishment of a model that assesses the ease or difficulty of transforming a given space over time. This approach facilitates the identification of the most suitable areas for such transformations (Adams 2007).

In general, these changes can be analysed within the following groups:

- Firstly, historical information can be collected using a typical data analysis methodology, such as a database that includes changes over time. This information can come from various sources, depending on the validity of the data we are studying. For example, if we want to reconstruct a large landscape, such as an entire region, we will need an overview. Agronomists and other geographical authors may have provided such perspectives, and their compilation is essential for understanding the entire process. However, this does not necessarily mean that they can be used as parameters for the analysis itself. Therefore, the first step is to gather the principles that can be analysed, serving as a guide to help us evaluate if there has been a significant change in the landscape, its level, and extent.

- This leads us to the second issue: once a change is detected, we must track it over time. Knowing its approximate location and magnitude, we need to understand how quickly it has occurred, whether it was a singular event or a long-term process. In the former case, we are primarily talking about catastrophic events, such as earthquakes or tsunamis that substantially alter a territory. In these cases, the closer we are to the event, the easier it is to study it. If it is the latter case, we are interested in understanding the evolutionary process, whether it is due to human or natural factors. That is why we need to rely on sources of information not only from Roman times but also from medieval and later periods to determine the extent of the change.

The lack of data in this regard should not be interpreted as the absence of change, as these changes may have gone unnoticed by contemporaries or simply were not recorded. Therefore, the best solution is to identify clear moments of change, that allow for comparison. For example, multiple authors describing the same river over time will inevitably include contradictions and variations, enabling us to make an overall assessment of long-term changes.

In the case of studying topography in a large area, as can be observed, the methodology is diverse but shares several common elements. Firstly, it involves identifying changes with significant milestones in time that provide certainty that these changes have occurred. Secondly, it requires having a map of the current topography that provides basic data for its reconstruction. Thirdly, it necessitates having quantifiable elements that can be used to reconstruct this topography based on the aforementioned criteria.

Conversely, when working in a very specific area, the main challenge is the quality of the data available for reconstruction and the possibility that historical base information may not be representative of the site. Only in cases where areas are exceptionally well documented by historical sources, such as the description of a major city, should such attempts be made. Nonetheless, achieving a finer reconstruction of a specific archaeological site or landscape is possible, especially for times other than the Roman era. In this context, we are primarily referring to a fundamental concept, which is a foundational proposal. The pre-industrial landscape is closer to the mechanical transformation of the countryside during the Roman era than it is to our current landscape. With this premise in mind, we can attempt to reconstruct what the topography was like several years ago, for which we have detailed information allowing us to visualize and reconstruct this data on the current terrain model. Methodologically, it is sufficient to georeference the old topographic lines or elevation curves and add or subtract them from the current terrain model.

In addition to topography restoration, this model can aid in making strategic decisions for surveying and prospecting the territory by identifying areas with a higher potential for archaeological sites.

From a methodological standpoint, these models can serve to assist in making strategic decisions for conducting territorial surveys and identifying areas with a higher potential for archaeological sites. This is a logical approach. If we understand that a particular area has undergone significant erosion, regardless of the existence of archaeological sites in the past, they may not be detectable. Conversely, in areas where deposition has been substantial, archaeological sites may be buried at great depths and not detectable on the surface. This information can be used to create viability maps for finding archaeological sites.

Furthermore, based on these types of studies, one could even consider other questions, such as the viability of archaeological sites by regions. For instance, in the case of prehistory, certain archaeological sites can be located on river terraces where favourable conditions existed at that time. This is ultimately a topographical criterion for identifying archaeological sites. Why can we not apply the same approach to other historical periods? The method proposed here would serve this purpose by defining a historical topography that can be used to reconstruct the living conditions of each specific culture, thus contributing to predictive models, which will be discussed in subsequent chapters.

6.1.1 Conditions

The conditions to be taken into account for the model can have two origins, natural or anthropogenic. Within the former, we can basically classify them into four main groups, as follows.

First, we have to consider geology as the main factor, as it would determine the possibility of changes over time. Generally, in areas where there has been a very large anthropic action, for example in urban areas, it is currently very difficult to see the change in topography. We may be lucky enough to have certain cartographies and historical data in greater detail, but geography will not normally have as much information in any case. Considering precisely these geological levels is essential for our study, since in a general analysis we can see that we can almost always find three different elements. First, the geological levels that we have seen would not be altered. Second, a space where we could not be sure because they are interfaces between consolidated areas and others with changes. In this case, we usually would need to further study the geology to see when it was created (Wilson and Gallant 2000). Third, an area where we are sure that in Roman times it would have been different, for example, alluvial deposits.

Once we know the geology, we must focus on an agent that can modify the environment, like the wind. It is the most common in all areas. This gives rise to wind erosion which will have little effect and yet there will be other areas where it has more impact, such as dune areas. These can be easily modified by wind displacing large quantities of sand, which means that we cannot know if that area was exactly like that at another time. On the contrary, there will be certain wind sediments, which may well not be the same. This is something that can be easily observed through geology and determining what these aeolian sediments are.

The third main condition is the alluvial sediments of the rivers themselves, the erosion that occurs by surface water but also in the deposition of sediments. The former is in the rivers themselves and we understand that they have been eroded, while the sediments will be seen in the change of the coast. This has to be complemented by specific studies so that we can look at how this river basin works, understand it better and know whether these depositions have been relatively more recent.

At the fourth level of analysis, we can have changes that are related to catastrophic events, and events at particular points in time that can modify the terrain. Of course, earthquakes themselves must be taken into account, which can substantially modify the structure of the Earth in certain cases. In this case, we can refer to the previous literature to establish whether there are any such events recorded. Moreover, especially from the point of view of coastal areas, events such as storms can rapidly modify the coastal space as can tsunamis, which can displace a lot of inland land masses. These are the main variations to support our understanding of the topographic edge of a territory.

In contrast to natural conditions, human activities have also the potential to significantly modify the topography. The lack of recorded data regarding anthropogenic modifications poses a significant challenge. While these modifications may be smaller in scale compared to catastrophic events or natural processes like sedimentation and shoreline changes, they are still highly relevant, particularly in specific case studies. This brings us once again to question the scale of our work. In large regional studies, anthropogenic modifications become significant, primarily in the context of major projects, such as the deliberate alteration of a natural area through dam construction or river channelling. However, when working at a more localized scale, it becomes imperative to consider these anthropogenic changes (Keane et al. 2018).

Examining historical cartography and other archival photographs can provide valuable insights into recent and long-term changes in the landscape, including increased sediment deposition. It is also important to consider anthropogenic actions, particularly in areas where deforestation has been extensive. Massive deforestation at a certain point in time, for example, can result in heightened erosion and subsequent alluvial and aeolian deposition, leading to the destruction and alteration of the topography.

When faced with significant environmental modifications, such as mining or stone exploitation, restoring the original topography becomes a challenging task. However, if such activities are ongoing, it may be of interest to reconstruct the original topography for the purpose of other modelling efforts. In such cases, having access to cartographic information becomes essential. Even if the available topographic data is not highly precise and dates back a century or so, it can still provide a valuable starting point for approximations and analysis.

As can be seen in Figure 8, the changes made in individual case studies are too large to be summarized. Fortunately, most of the criteria can be modelled by taking into account the actors involved and assessing the level of involvement of each one. For example, when we look at an archaeological site, we can assess the state of conservation. We would do something similar but with other natural elements in order to know the possible transformation of the landscape.

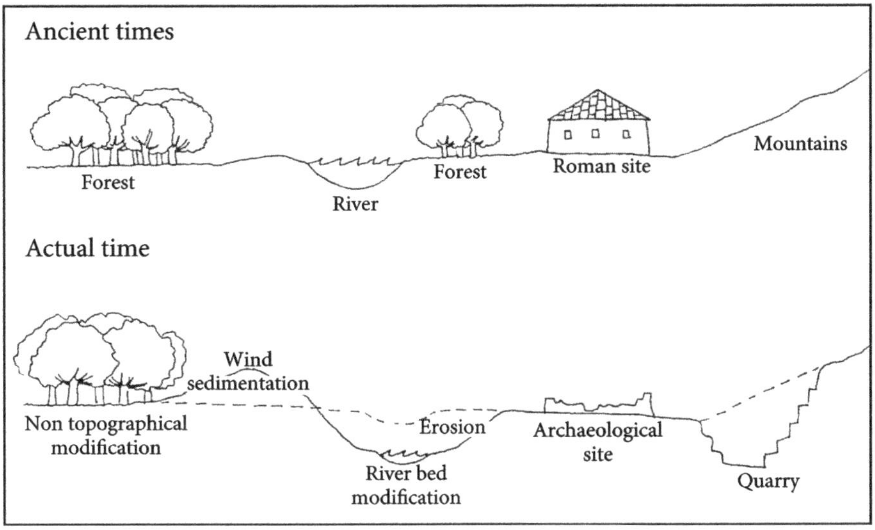

Figure 8 Comparison of changes in landscape topography between ancient times and nowadays (source: E. Toscano).

6.1.2 Method

From a methodological perspective, the approach to assessing topographic changes differs between large study areas and specific case studies (Figure 9). When examining a specific case, such as a *villa*, the topography can be evaluated by considering the typology of the terrain and understanding the historical and geological levels (Petrasova et al. 2018). Factors such as different land use or agricultural practices may have influenced the topography in these cases.

The proposed model incorporates three key constraints: topography, geology, and historical mapping, which together enable the reconstruction of a given space. Geology plays a crucial role in assigning values to areas that are more likely to have undergone changes. This is achieved through reclassification, where suitability values are assigned based on our understanding of the potential for modification.

In conjunction with geology, the current topography of the area under study is considered. To accurately reconstruct heavily modified areas, a high-resolution digital model is essential. By analysing the topography, we can identify areas that have experienced significant changes. These changed areas become the focus of our intervention and restoration efforts.

Historical cartography plays a fundamental role in the process of restitution. The accuracy and detail of historical maps are crucial, as they serve as references for how the area appeared in the past. For instance, if a river's course has been altered, we can partially restore it by georeferencing a historical map that depicts its previous path. While challenges may arise during the restitution process, the goal is to achieve the highest level of accuracy possible.

In cases where modifications are recent and precise topographic data is available, such as in the case of a quarry, the model can be refined to produce highly accurate reconstructions. The combination of topography, geology, and historical cartography provides a comprehensive framework for the reconstruction of modified spaces.

For small-scale examples, geology can provide valuable insights. One approach is to reclassify geological layers and assign values based on the potential for modification. This method is more complex and restitution may be challenging, but it allows for the establishment of different models.

Another approach involves using a digital elevation model of the current terrain and selectively cutting out areas to approximate the topography at a different time. For instance, if archaeological evidence suggests that the geology is 2 metres below an alluvial space, which may correspond to the Roman period, this height can be established.

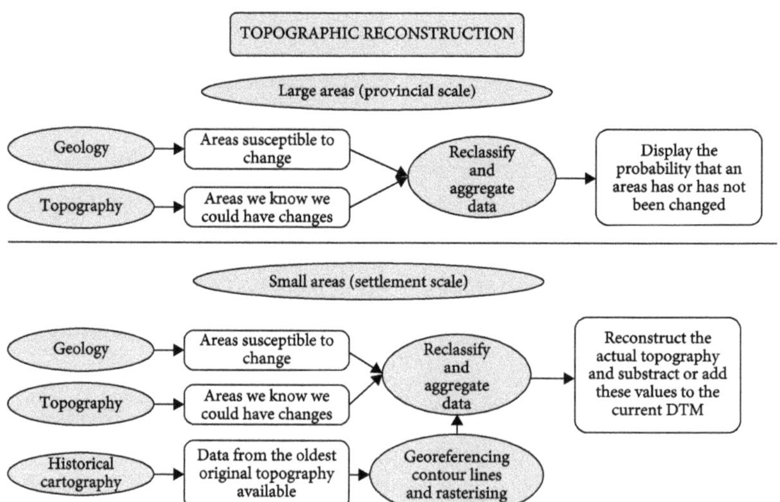

Figure 9 Scheme for reconstructing the ancient topography in large and small areas (source: author).

By interpolating this information across the study area, it becomes possible to identify the areas that may have undergone modifications. These approximations can help determine specific locations to investigate for potential archaeological sites. Areas that have been significantly modified may show no archaeological evidence due to devastation or sediment accumulation. Conversely, sites may be preserved in areas that have experienced elevation changes. The study of archaeological sites and their corresponding elevations can complement this information.

6.1.3 Application Examples

The model we present here is a complex and challenging approach with limited parallels. Rather than offering a defined method, we aim to provide different ways of approaching the problem. We are currently collaborating with colleagues from the University of Almeria, Spain, on several studies focused on reconstructing ground conditions in an area that has undergone significant and rapid changes.

These studies, address the technical complexities associated with studying topography in a complex landscape. The area under investigation experiences intensive deforestation and seasonal but torrential rainfall, leading to pronounced erosive processes (Aragón et al. 2024).

In addition to our work, there have been notable case studies where the classification of the territory has been employed to reconstruct historical topography. An excellent example is the work of Boak and Turner (2015), who reconstructed ancient coastlines. While historical approaches to this problem date back to ancient times (Desmet and Govers 1996), recent studies primarily focus on classifying coastlines rather than achieving complete restitution. This perspective is valuable as it seeks to identify areas that have undergone significant modification over time rather than striving for exact reconstruction (Iwahashi and Pike 2007).

Overall, the approach to studying historical topography is a complex and evolving field, involving various techniques and methodologies. Our ongoing research and the works of other scholars contribute to expanding our understanding of how the terrain has changed over time and provide valuable insights into the classification and analysis of modified landscapes.

In addition to this, it may seem that this topic deviates from the main focus of the book, as it appears to be a matter without a historical question. In reality, it is quite the opposite, as the ability to reconstruct topography is merely the first step in understanding a landscape or its changes, which have significant cultural components. In our case, restoration is a necessity whenever there is suspicion of a change in the landscape from the present, making it difficult to predict. We are currently working on a case study still in progress where the change in topography at a specific archaeological site is being debated. The significance of this data lies in the existence of a Roman reservoir, so restoring its area and capacity is crucial to understanding its use, whether for water supply or irrigation. The methodology in this case is similar to what has been previously discussed.

6.2 Reconstruction of Topography for River and Maritime Navigability

Navigability as a variant of this study of historical restitution of topography can provide an understanding of the navigable capacity of a certain space. Here it is a specific case of study that derives from the previous one, but that has very specific conditions that can be analysed in certain circumstances. We know that certain rivers have the possibility of being navigable because we have historical information on a series of rivers whose conditions in Roman times may have been navigable, but we do not know exactly to what extent. Either the river has had certain recent modifications, especially due to the issue of dams and the position of the river, or the ancient conditions are unknown. For the Roman period, the reality is that to determine river navigability, it is necessary to know a series of historical and conditioning factors (Hall 2008).

The navigability of large rivers has been extensively studied and is unquestionable. The challenge arises when comparing rivers of different scales. Specifically, intermediate-scale rivers, such as those found in the Iberian Peninsula, are more difficult to assess in terms of their navigability. Although this study has a diachronic nature, we will primarily focus on the Roman period, as it provides us with sources that inform us about the navigability or non-navigability of these spaces. The continuity of their navigability throughout history will be determined by changes affecting their natural or anthropic configuration.

In terms of historiography, rivers have been regarded as transportation routes for people and goods, on par with the sea, but mostly in the case of large rivers like the Nile. This river serves as a reference point for ancient times. However, determining general navigability conditions is relatively straightforward, so we begin with the assumption that any river meeting these conditions would have been used for transportation. Estimating these conditions is an integral part of our study. The first challenge lies in historiography, as there is a dearth of studies on this topic. In other words, aside from major cities, there is no direct correlation between the presence of cities along rivers and their utilization as navigable routes. Assessing the navigability conditions of a river for a specific historical period is not difficult, as the draught of a ship is the most crucial factor.

6.2.1 Conditions

The first of the determining factors is the historical cartography and documentation itself since if a river was navigable at a given moment in its history, it could have been navigable at earlier times. Here we are also talking about a series of geomorphological changes such as those we have seen previously when defining, for example, the coastline and making a very dynamic space (Weiss 2000). These transformations can be very large, mainly a constant flow of water at a given time with sufficient draught to allow a certain boat (Villalba 2018). In those cases, we could differentiate two major options. On the one hand where the river could be navigable by the depth of the river basin. On the other, a river is navigable by another factor, like the rise of tides.

When working with the Roman period on issues like coastline changes, it is essential to exercise caution and ensure the reliability of the data. There is ongoing debate about whether sea levels were higher or lower and whether the coastline was inland or further out. The prevailing understanding suggests that the coastline itself was generally at a similar level, but the processes of deposition and erosion make it a highly dynamic space. Cliffs, despite appearing degraded, may still preserve fossil remains, while river beaches can vary in size or may not exist at all. River mouths, in particular, exhibit dynamic characteristics, forming deltas and estuaries (De Paz et al. 2006). Human activities have also significantly modified these areas, and studying these anthropic influences can be challenging without historical information, such as records from medieval or modern times. The precise position of riverbeds can be altered by factors like meander formation or other anthropic events, which may have partially obscured certain locations (Trapero and Aragón 2022).

In both cases, it is crucial to consider the relevant conditioning factors, such as tides and river basins, to understand the historical landscape. The first factor, tides, can vary in different areas and significantly affect navigability. For example, in the Atlantic Ocean, tides can rise and fall by up to 2 metres, unlike the Mediterranean. Additionally, coastal conditions, including the formation of sand bars, deltas, estuaries, and other geomorphological features, can either facilitate or impede navigation. Areas that currently have delta formations, for instance, may have multiple channels that were once navigable under specific circumstances (Trapero 2019).

To assess these conditions, it is necessary to consult existing geomorphological studies of the coastline, along with available data on navigability, particularly in areas with strong tides. Archaeological evidence also plays a vital role, as certain sites, such as salting factories located on the coastline, provide valuable insights into historical navigable areas. Additionally, geological studies or sampling can provide further information about areas that have undergone sedimentation or are known to be filled.

Regarding river analysis, it is essential to understand the hydraulic potential of a river, which can be calculated using hydrological tools, as mentioned earlier. This involves considering various conditioning factors such as rainfall patterns, underground filtration, and evaporation. The goal is to reconstruct the maximum extent of a river's basin, taking into account its potential reach and hydraulic characteristics (Figure 10).

6.2.2 Method

The methodology utilized to assess river navigability follows a relatively straightforward approach. It predominantly involves the selection of rivers potentially navigable during the Roman era, with considerations extending to

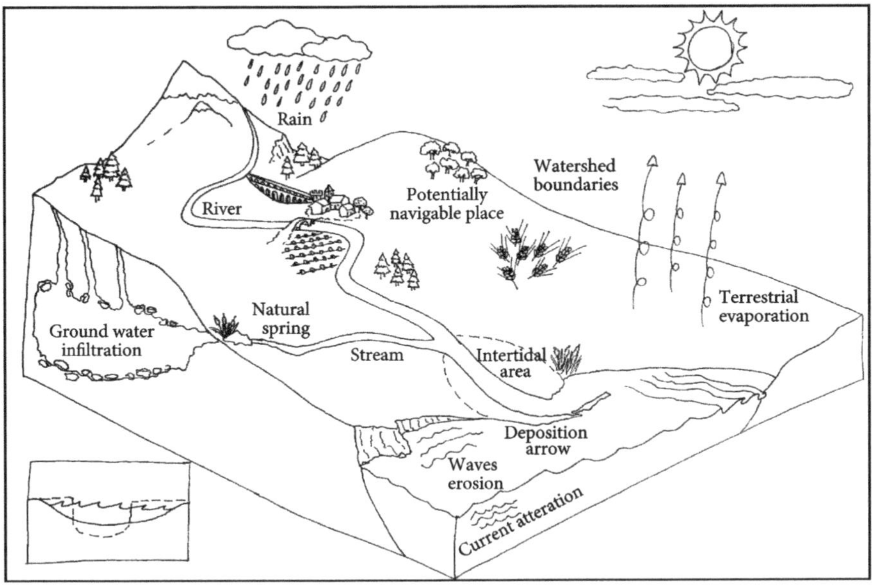

Figure 10 Scheme of water circuit with the main criteria used in the model (source: E. Toscano).

subsequent historical periods. Once pertinent conditions are compiled, a model is meticulously crafted to reconstruct these variables. A pivotal determinant of a vessel's ability to remain buoyant is its specific draught, intricately influenced by both the materials and shape of its hull, as well as the nature of its cargo. For instance, even in shallow waterways, navigation remains feasible with a raft designed to float on the water's surface, demonstrating the adaptability of vessels to diverse navigational conditions.

There are two main scenarios to consider (Figure 11). The first scenario involves navigability due to coastal conditions, necessitating an examination of changes at the river mouth. This analysis encompasses understanding transformations such as the formation of coastal arrows, deltas, or estuaries. Topography significantly influences this analysis, as the elevation generally increases inland, and changes in tidal influence or river topography tend to be minor. Consequently, our focus is limited to a specific area near the river mouth.

The second scenario is more complex. Here, the objective is to calculate accumulation based on the estimated flow potential of the region using the flow accumulation tool. This tool provides information about the number of cells or pixels that contribute to water discharge in a particular direction. At the river mouth, we have the total number of contributing cells. The key aspect here is the ability to simulate the hydraulic potential of the river, which is determined by the volume of water that would flow at that specific point. By obtaining this value, it becomes possible to calculate and distribute flow over time, ultimately measuring the riverbed and draught. This data is crucial for understanding the hydraulic potential of a river within its basin. In this context, we assume that the present conditions resemble the past more closely than undergoing drastic changes, particularly when considering long-term regional comparisons.

The methodology employed is both experimental and theoretical due to the lack of precise information regarding precipitation climatology, infiltration levels, and water input from underground aquifers. To compensate for this, we measure the minimum potential flow level, assuming constant rainfall throughout the year. Additionally, we assume that 9 out of 10 parts either evaporate or contribute to aquifers, with only one part circulating in the river. These assumptions represent particularly unfavourable conditions for navigation. The final value is converted into flow (m^3/s), and subsequently, the channel area is calculated, assuming a semi-circle or ovoid shape to determine the potential draught.

Through this approximation, we obtain a minimum flow value. If this value exceeds 1 metre of draught, then the river is deemed navigable at that specific point. To calculate this, we model the riverbed's depth and width. It is important to note that the riverbed can undergo significant changes over time. By simulating the shape of the river as a semi-circle or oval, the calculated values may not be exact but provide a range of options that can be more or less navigable. This approach helps us understand the sufficiency of draught and the expected river profile at that point.

However, if even with this minimum calculation, the value is significantly lower, it indicates that under more favourable conditions, the river would still not be navigable. In such cases, neither higher and more consistent rainfall throughout the year nor an estimate allowing for navigability during half of the year would make a substantial difference.

Therefore, the more hydrological information available for the area, the better the methodology for reconstructing the hydrological conditions. Hydrological calculation tools are employed to determine the potential flow that would pass through a given point when rainfall capacity is unknown

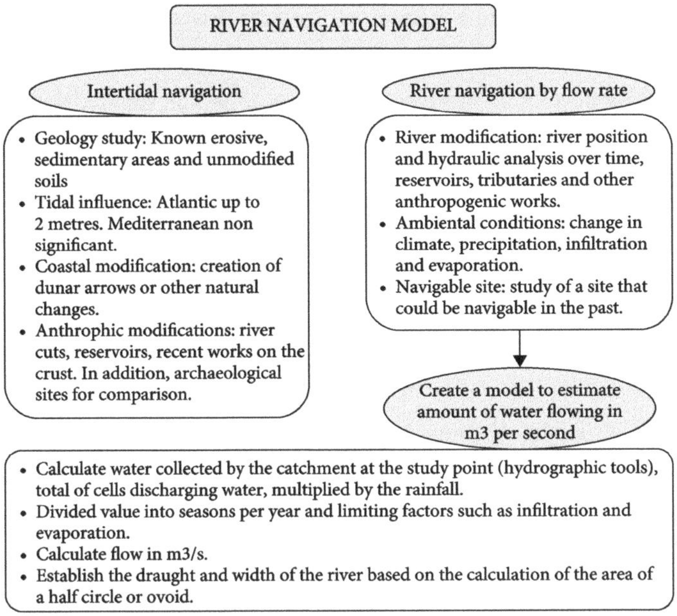

Figure 11 Workflow for creating a river navigation model with intertidal and flow rate factors (source: author).

(Khatami and Khazaei 2014). The aim is not to determine a positive answer, but rather to qualify a negative one, recognizing whether the river is likely to be navigable based on the modelled best conditions.

6.2.3 Application Examples

As a case study, we present an article that models these conditions specifically in the Guadalete River (Trapero et al. 2023b). In this case, navigability has been reconstructed for the medieval period, as there was specific documentation on this problem. Specifically, the difficulty of reaching the port of El Portal, which was navigable but with shallow areas where ships ran aground. After analysing the conditioning factors and reconstructing the flow potential, it is clear that the main agent favouring navigation was always the tide, not so much the potential of the river itself. Likewise, geomorphological changes were not as important as anthropogenic changes in the change of navigable conditions. We are currently experimenting with this perspective through several case studies, such as the one mentioned above.

Extensive literature and historical sources provide insights into these navigational difficulties, as well as evidence of river modifications made to facilitate navigation. In most cases, navigation relied on the hydraulic potential of the river itself and the influence of tides, and we can model both. Apart from this specific case study already shown, there are currently limited examples in the literature that utilize this methodology. However, we believe that studying these conditioning factors represents a promising avenue for future research. The scarcity of studies on modelling these factors may be attributed to a tendency to view rivers primarily as barriers rather than as means of communication. Additionally, the challenge of modelling large and dynamic spaces like river basins throughout history further contributes to the limited research in this area.

We do not clearly find other case studies that address this issue, but we are aware that it is an active idea pursued by other research groups, such as the calculation of vegetation and potential topography to assess these changes (Andresen 2007). Sharing these approaches is important for modelling such questions collectively. The concept of calculating navigability is another application related to landscape change. Its applicability is entirely cross-cutting and diachronic, allowing us to work on historical periods other than our own.

Although not identical, there are other approaches in other periods that allow for the understanding of a significant portion of this landscape change. This is particularly evident in prehistoric periods where changes in topography can be highly significant. Therefore, calculating passage zones such as channels or similar features can be important for proposing subsequent models such as mobility patterns (Murrieta 2012).

7

Mobility Models

Mobility is the movement of living beings and objects in space. One of the most interesting analyses we can do with GIS is the calculation of optimal routes between two spaces. How long would it take to get from my house to the nearest supermarket? A simple example that we are currently calculating with Google Maps.

Mobility has significant applications in many aspects of our lives, from understanding routes and timings, which we already have today in delivery services and courier companies with order tracking platforms. However, this is just the tip of the iceberg, as the real potential lies in route management and administration to ensure the effective functioning of a delivery company. When applied to history, it serves to comprehend where people are located and how they move across the territory, depending on the mode of transportation they use, their own mindset, and the available resources.

If we want to do this analysis for the past, we need to know a set of spatial and technological conditions with tools that allow us to understand them. Among the spatial analysis tools, there are different conditions and vehicles that allow great viability of options. This means that using a current tool is not possible, but that we have to modify it to meet the old mentality and landscape criteria. In contrast to the models expressed in the previous chapter, the tools are well treated and there is an extensive bibliography.

The tools for making these models are many and relatively complex. Therefore, we will mainly talk about two of them: distance costing and route costing. The first one calculates the cost of time, energy or whatever we want from one point to the whole territory; the second one relates this cost to a specific place by creating a route. The simplest process is to generate through the given constraints, which can be any raster, a distance cost map, or isolines showing the cost from the given point, line or polygon of origin. In addition, we usually generate a second map, called a backlink in which we have, for each cell, the direction in which the theoretical object would move. With these two maps, you can use another tool such as the path distance, in which we include other entities to calculate the optimal paths between them. So far, so good, but nothing is further from reality. Mobility analyses are very complex since we are simulating how the movement on the ground was at a given time, with a particular technology and vehicle, and therefore, there are probably substantial differences. We will explain all these points in detail.

In general, we must first understand that to move any object there are forces in favour of the direction of movement and others working against it. Since the model is going to estimate those in favour, we have to include the others. That is why, to make a competent mobility analysis, it is necessary to create friction maps, which are raster layers, like the pixels of an image, where we say how difficult it is to move depending on an algorithm that converts these values into numbers that we can understand, usually time or energy (Herzog 2013). In the case of the terrestrial environment, it can be the slope. Creating a friction map consistent with the historical question is basic to generating a correct model. For this, we have to make the calculation using a friction algorithm that converts this theoretical cost into terms we can understand, usually in time or energy (Marquez et al. 2015). There are several approaches to this subject (Tobler 1993; Llobera and Sluckin 2007; Baek and Choi 2017).

Currently, it is possible to establish a friction map using the slopes of the terrain, plus a series of mobility limiters, in the case of crossing a wooded area. It is also possible to create places of transit preference, as in the case of a network of established roads. These two criteria are very complex to apply to historical times, especially the further back in time we go. We often do not know how the landscape has changed or if an area that is impassable today was impassable in the past and vice versa. That is why we must know the territory we want to work in very well in order to create conditions that fit reality.

To create a friction map we generally start from the slopes of the terrain. We can calculate from a point, line or polygon in space the cost of moving to contiguous cells and the last cell of the raster. This calculation is made based on the friction we give and is consistent with the direction (it is easier to go down a gentle slope than up it), for the case of the isotropic method to calculate optimal routes. There is another option called anisotropic that takes into account the outward and return journeys, so it considers both uphill and downhill terrain (Herzog 2014; White 2015). The main application of this is another raster that visually shows us which areas are more accessible and which are not.

As mentioned before, various limiters or enablers can be applied to this friction map, which in essence modify this layer of slopes to increase, decrease, add or remove areas from the analysis. This is useful when we know that there are constraints that are not the real ones, and we have to change this friction layer. For example, rivers and streams should be taken into account, as normal mobility does not understand that there is water, and we need to tell the tool that this flat area without obstacles cannot be traversed.

In general, mobility modelling is common practice with many recent approaches (Howey 2011; Verbrugghe et al. 2017; Alberti 2018; Carreras et al. 2019; Milheira et al. 2019; Verhagen et al. 2019). Most of the analysed approaches use two complementary tools, namely path costing and path generation optimization. This is the same as using GPS on a mobile phone or in a car to determine a route. The spatial constraints can be very different depending on the type of vehicle we are considering and the terrain. So, the first thing to do is to define what is moving, how it moves and where it moves, as well as any obstacles or constraints to movement that may have existed in the past, and then examine the current capabilities we have to model these possibilities with GIS. Undoubtedly, our discussions primarily revolve around conditions within the terrestrial environment, encompassing the movement of individuals and animals across the land. However, in ancient times, we encounter other modes of transportation, such as ships navigating rivers and seas, which considerably complicate the task of understanding both the internal dynamics governing a specific vessel's movement and the external conditions associated with water bodies.

The models shown here attempt to understand the issue of mobility from the ancient viewpoint and the physical constraints that existed. For this reason, we have divided this chapter into four parts. The first is on the generally considered terrestrial model, which has been well studied. The second is on the variation of river navigation and how to deal with these elements. The third considers maritime mobility as another form of transport. In addition, the final part looks at the economic costs of transport and the possibility of doing studies of this type.

7.1 Land Transport

It is important to distinguish between mobility analysis and modelling. The aim is to understand human behaviour in relation to physical factors, rather than simply calculating the most efficient routes. It is crucial to grasp the constraints, adjustments to the tool, and possibilities involved. Rather than solely focusing on the optimal way to reach a destination, the objective should be to comprehend the movement within a territory. This entails determining the proximity or distance between places and identifying natural pathways.

In other words, calculating the best route based on current parameters is not significant for the past unless we take into account all the mental background and specific conditions to ensure that the results correspond to a particular mindset and timeframe.

A simple approach to understanding mobility is to consider the routes that individuals would take. If a person starts walking in the middle of the countryside, not knowing the destination but following some guidance, he or she will use the logic of taking the easiest route. If the place of arrival is known perfectly well, then the path is more certain, but the problem is the same. With costing tools, what we do is to calculate the best way to go to or from a place. It is the same as if we are walking along a road and see an obstacle and turn around. We determine that the cost of this stretch is too high, and it is better to make a detour. We do something like this for each cell of the raster we are studying.

Things get more complicated when it is not a person but another vehicle. This is primarily because a person today and a person from two thousand years ago may be similar, while the parameters for modelling this with a pack animal or a vehicle like a cart are more uncertain.

For example, walking through a mountainous area with dense vegetation and steep slopes, or crossing a swampy terrain with a heavy load, presents different conditions compared to a flat road. Mobility, whether on foot or on horseback, requires not only identifying the best route but also understanding the distances between points. This can be particularly interesting when considering distances and landmarks along main roads, which often correspond to specific archaeological sites or stopping points separated by a certain number of travel days.

A terrestrial mobility model encompasses numerous options and variables, each with its own conditioning factors. Transporting a person is not the same as transporting an animal or having an animal pull a cart. Even among different types of animals, there are subtle differences.

The main criteria for studying mobility can be categorized into internal and external factors. The internal criteria pertain to the individuals, animals, or vehicles involved, while the external criteria relate to the surrounding environment. From the perspective of ancient mentality, the study of internal criteria is largely comparable to that of the present, with some considerations for technological differences, if applicable. However, the models we develop for different types of moving objects can generally transcend time.

The environment in which movement occurs is a distinct consideration. In broad-scale studies, the general topography remains relatively unchanged, posing minimal issues. However, in smaller areas, the significance of topography may vary. Additionally, factors related to societal mentality play a crucial role in understanding mobility, what drives movement and destination choices. While our focus often centres on the transportation of people or heavy goods, we must approach the analysis with caution. The assumption that the shortest route is always the preferred one may not hold true in ancient mentality. These aspects will be further explored in the final subchapter dedicated to mobility models.

On a technical level, as already mentioned, it is best to use two tools, distance cost and path distance. With both, it is possible to do almost any kind of analysis at least on the ground. One specific approach for mobility analysis is the MADO analysis (Fábrega and Parcero 2007), which employs hydrological tools to simulate watersheds. Various variables can be incorporated, each with its own configuration, but all can be created using a fictional map as the foundation for this type of study.

7.1.1 Conditions

For the conditions, we need to understand different aspects, such as the vehicle itself and the constraints of the terrain, and internal and external conditions. The main 'vehicle' for human mobility is their own legs. A human being has several basic limitations that can be modelled in a GIS. The ability to walk or run is limited to some conditions, but on average, an adult can walk at a comfortable pace at about five kilometres per hour (km/h). For such a calculation, we do not usually consider running, as it does not allow long distances to be achieved, except in exceptional cases. It is also usually analysed as a continuous movement, i.e. there are no stops or sleep breaks (Collischonn and Pilar 2000; Sarmady et al. 2014).

The main factor is the slope of the terrain in large areas, combined with more specific factors such as obstacles, whether they are areas of difficult movement such as crossing a river or easy passage, like a road.

For a historical model, a human being today and in the past has the same constraints, but this is not the case for vehicles. Nowadays, we are used to travelling at 120 km/h by car or flying at over 800 km/h. We know that in the past, the driving force of mobility used to be transport animals, such as horses, donkeys, mules and oxen or their combinations in carts and wagons (Haisman and Goldman 1974; Herzog 2014; Fonte et al. 2017; Carreras et al. 2019). The volume and weight of the load to be transported will also greatly modify this mobility (Burford 1960).

Transport would preferably be carried out using bullock carts. We know from Roman legislation that they could not carry more than 450 kilograms (Carreras 1994). This is an estimate based on the law, although their capacity could be much higher, even more than a tonne, 1,000 kilograms (Burford 1960; Raepsaet 2002).

This is a controversial issue for two reasons. Firstly, if the legal limit is stipulated in legislation, it is probably because it was constantly exceeded. Secondly, the logic of transportation does not allow for including excessively heavy loads due to the pulling force of animals and the preservation of the vehicle, which would be seriously damaged by heavy weights. In any case, the issue could be resolved with vehicles adapted to specific needs and by adding more draft animals. For us, the current interest lies in the fact that a cart loaded with a weight exceeding 450 kilograms represents a significant burden on the animals, which alters their overall performance.

An ox-drawn carriage can travel an average of 25–30 kilometres in a day, at approximately 1.14 km/hour (a mile is approx. 1.5 kilometres). According to Elius Aristides (*Theod.*, 27, 1–8), a journey of 42 miles can be made in a day without rest, but if rest were taken it would be approximately 1.16 km/hour.

Horace (Sat. 1.5) describes a journey of an average of one mile per hour. In general, we could establish an average chariot speed of 1.6 km/h.

Whether it is a person, animal or cart that we want to study, we have to know the place of origin and, if possible, the destination, the type of movement and the difficulties encountered along the way. For example, in our model, it is very different to calculate the mobility of a single person, an army, and then a wagon loaded with several tonnes of weight. On land, the general conditioning factor is the relief of the terrain itself, since it is easier to travel on the flat than on a mountain. The difference in slope is what favours or hinders the passage, something we can easily calculate with a Digital Elevation Model. For example, in the case of a person walking on flat terrain, we would go at a good pace, while if it is steep, we would see our speed substantially limited. In addition, there are other constraints, for example, areas where it is difficult to walk, such as on waterlogged ground or swamp or overgrown vegetation (Grau 2011).

In a current analysis, the constraints related to mobility can be assessed using spatial data in a GIS (Figure 12). However, when studying past epochs, we must first estimate these constraints. Generally, both animals and people find it easier to move on flat terrain with a slight slope, while going downhill and uphill takes more time and energy (Lawton 2004). The difference lies in the varying time required and the distance that can be covered by different modes of transportation. For example, a person or animal can walk for many hours a day, as seen in pilgrimages like the Camino de Santiago. However, their movement is typically limited to a certain period per day, coinciding with daylight hours that change with the seasons.

The topography of the land, during historical times, has not undergone significant transformations, although this depends on the scale of analysis. At a large scale, for example, which involves micro surveys in villages near the coast, the accuracy and resolution of the data may be more questionable. Conversely, at a small scale, when examining mobility between cities hundreds of kilometres apart, changes in the territory and landscape are easier to determine, and the margin of error is likely negligible. Several applications have been developed in this field, such as the freely available online tool ORBIS, which demonstrates the combination of these constraints at a small scale, particularly in the Mediterranean region (Scheidel and Meeks 2012).

Among all the conditioning factors, one aspect of particular importance in this type of analysis is the presence of bodies of water. Rivers, streams, and lakes have undergone changes over the historical period, not so much in their overall basin, but in terms of their size and position. However, crossing these water bodies can be challenging unless anthropic structures like bridges have been constructed. Therefore, it is crucial to have knowledge of these human-made conditioning factors.

Therefore, to generate a realistic terrestrial mobility model, we need sufficient information about internal and external factors. Internal factors will vary depending on the type of vehicle we are using. External factors are mobility constraints. As we will almost always perform calculations in large areas to determine routes between cities or similar locations, the main constraints are slope and obstacles such as rivers, coastline, etc. Other factors that can be included are geology, for example, soil conditions that affect ease of transit or wooded areas. However, these factors depend on the level of precision of the data and historical knowledge, so they can only be used in very specific cases.

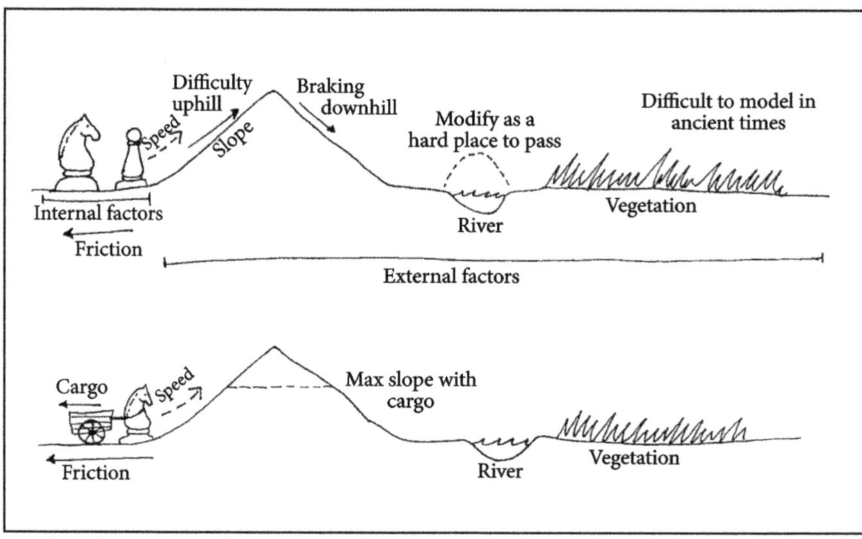

Figure 12 Comparison between the factors for a person's movement or a horse's movement with a cart. Some external factors such as the slope or the difficulty of crossing a river can be modelled, but others like the vegetation are hard for historical models (source: E. Toscano).

7.1.2 Method

As we have seen in the previous point, the technical difficulty of the study of mobility requires a basic understanding and a deep historical knowledge of what we want to investigate. Generally, the new practices that have been developed improve and complicate in many cases the precepts we have indicated above. However, the key lies in the conceptualization of the historical model we want to investigate. If we stop to think about the historical conjuncture that interests us, establishing the criteria, knowing the mindset and how the landscape has changed, it is possible to carry out a much more reliable GIS analysis that provides historical information. The technical problem of how to do this comes later, and nowadays there are enough tools available to be able to solve almost everything before programming.

Methodologically, on the one hand, we are going to explain how this would be done in a traditional way, through distance and path costing, while on the other hand, we will explain the MADO process. Both share many of the same tools and processes, so they are summarized together (Figure 13).

The traditional way to model optimal paths is based on using tools like cost distance and distance path in ArcGIS Pro. We briefly explained at the beginning of this chapter that these tools are the most important for calculating routes. With the first tool, we create rings in the cost factor we have defined (for example, hours or kilocalories if it is time or energy), so we know how much effort it takes to travel from an initial location to the rest of the territory. With the second tool, we calculate the optimal route between one or more starting points and other points, resulting in lines.

However, the difficulty does not lie in using these tools themselves but in creating and understanding what internal and external factors we are going to use. For external factors, it involves creating an appropriate friction map, while for internal factors, it requires an algorithm that transforms the cost into something measurable and corresponds to the specific vehicle in question.

Regarding the algorithm in question, we have already mentioned some examples that can be applied, which will be highlighted in the following section of application examples (Tobler 1993; Llobera and Sluckin 2007; Baek and Choi 2017). In general, these algorithms are designed for pedestrian terrestrial mobility, and it is currently challenging to obtain optimal models for other types of mobility.

In any case, the key lies in creating a friction map. As can be seen in Figure 13, the combination of elements we must consider for the friction raster starts with a slope map. Slope is the primary factor considered in the algorithms because it accounts for both positive and negative variations. However, to adapt it to historical constraints, we need to add other limiting factors. The way to do this is similar to other cases already explained. For example, limiting mobility by a river, treating it as a barrier to cross, is done simply by converting that line into a polygonal area, using a tool like buffer. Then, by converting it into a raster and reclassifying it, we can assign a value, for example, 10, to the river and 0 to the rest of the study area. Finally, we add our raster to the slope layer. This means that where there is no river, the value remains the same, while where the river is located, the slope increases by 10 degrees. Conceptually, it is as if there were a steep plateau where these bodies of water exist, making mobility difficult. This way, when conducting the analysis, the model will consider that bodies of water are not flat and easy to traverse but should be avoided if possible.

While the path analysis discussed earlier focuses on determining the best path between two entities, the MADO model allows for the identification of optimal corridors and connections as a network (Fábrega and Parcero 2007; Llobera et al. 2011; Parcero et al. 2019). The approach involves generating a potential mobility surface by applying hydrological analysis to a cost-distance raster, as already explained. This method utilizes friction rasters and hydrological analysis techniques to determine the most accessible paths, with the starting point being the source of a theoretical river or stream that flows through the areas.

The MADO model demonstrates a solid and accessible tool for addressing methodological challenges and holds great potential. Multiple analyses can be combined to create a theoretical communication network that encompasses various locations. By aggregating these layers, we can approximate the secondary paths within the territory. It is important to note that while the focus is on the MADO model, the crucial part of the process lies in the creation of the friction map.

As explained previously, the general process for conducting mobility calculations has been outlined. However, an additional precautionary step should be mentioned. Inaccurately defining the limits of rivers and streams can be a significant factor in producing unreliable results. The tool tends to prioritize flat areas, such as rivers and streams, when determining paths. Therefore, it is advisable to initially create a model using a raw raster that only considers terrain elevations to calculate a distance cost. This allows us to assess the optimal paths along the main rivers and streams. Uncorrected, cost analyses will presumably follow the main watercourses. While this additional step may take some time, it makes a substantial difference when compared to well-processed data. If the difference is minimal, it suggests that the model may not have adequately accounted for this constraint, requiring a re-evaluation. This is often the case in very wide river basins because even if the flow is limited, the flood basin can still be passable and therefore optimal according to the calculation. Whether this is true for the ground conditions or not depends on the researcher.

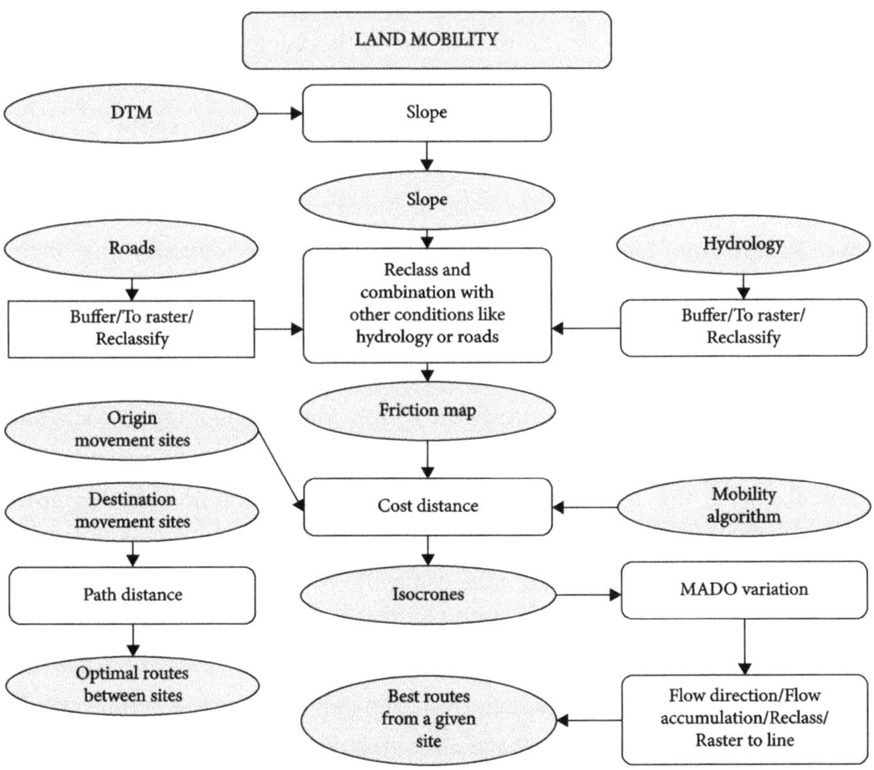

Figure 13 Scheme of land mobility model, with two variations depending on using optima routes between sites or best routes from a given site (source: author).

This simple control measure helps determine if the conditioning factor has been correctly defined. It should be noted that there is no ideal value for this factor as it depends on the specific terrain. For instance, in a highly mountainous area where main roads follow natural paths, following a river may not significantly impact the results since other conditioning factors are more influential. However, in a flat area, even a small stream can significantly alter the optimal paths. Further details on these considerations will be discussed in the following section.

7.1.3 Application Examples

Numerous case studies have been conducted using this type of method (Fábrega and Parcero 2007; Verhagen et al. 2013; Verbrugghe et al. 2017; Frachetti 2017). These studies primarily aim to reconstruct historical communications and understand mobility patterns, particularly in regions where passage holds geostrategic significance. As outlined in the methodology, there is a wide range of possibilities for modifying and adapting these approaches, which often serve as theoretical frameworks for understanding a more complex reality.

For instance, we applied this method to analyse a Roman city, such as the colony of Hasta Regia, where we generated a network connecting the various *villas* within the territory. The resulting network showcases the main connecting roads between these points (Trapero 2019). Other notable case studies can be found in publications that combine network and mobility analysis, such as those focusing on the Dacia region (Oltean and Fonte 2020) or modelling general communications within the Roman Empire (Scheidel 2014). The popularity of this method has been increasing in recent years, particularly in the Iberian Peninsula, where there are notable examples (Moreno 2022), as well as in other regions. It is worth highlighting the work on the hierarchy of archaeological sites using the MADO model and mobility analysis, including its application to Crete, Greece (Déderix 2016; 2017).

We have already mentioned that there are numerous case studies since this is a tool with tremendous potential. Describing all relevant case studies would require an entire book, so we have chosen some of the most notable ones. Specifically, we have selected the case of MADO because it provides a novel perspective on this type of analysis.

These types of models are becoming increasingly complex over time, facilitated by the vast corpus of previous cases and computational examples, thus enabling more accurate approximations. For instance, a recent advancement involves the proposal of a distance cost method that incorporates potential visibility within it, essentially performing an inverse Minimum Angle of Detection (MADO) on its position. This approach aids in understanding movement towards a prominent location, considering whether being seen or not serves as a reference point. This will be further elucidated in the subsequent chapter (Quirós et al. 2024).

However, when applying these principles, caution must be exercised on two fronts. Firstly, the accessibility of the tool enables many researchers to apply it easily. While this accessibility is not inherently problematic, it can sometimes lead to unreliable results, particularly if the friction raster is not constructed accurately. Consequently, a significant percentage of analyses in this regard lack validity.

Secondly, historical inquiries typically concentrate on establishing connections between two spaces. While this approach is legitimate, particularly when seeking to comprehend a road as a Roman thoroughfare, there exist numerous other analytical possibilities within similar areas. For instance, the determination of the minimum living area or the exploration of resource utilization around a *villa*, as discussed in earlier chapters, presents alternative avenues for investigation. Moreover, these principles can be applied to other spatial contexts, such as navigable routes or the incremental cost of transportation, as will be discussed presently.

7.2 Rivers and Streams Constrain and Support Mobility

In the ancient world, terrestrial routes were commonly utilized, but there were also secondary routes that could even surpass them in importance, such as river navigation. Whenever feasible, the transportation of goods and people through navigable waterways was highly likely. In path distance analysis, we typically consider land routes, but we must recognize that they may incur higher economic costs and have lower carrying capacities. Hence, the fact that main cities were often situated near the coast or navigable rivers is not a speculative notion, but one supported by historical sources and archaeological evidence. Proximity to water bodies was primarily driven by industrial waste disposal, and transportation, and not necessarily for water supply, which was often facilitated through wells, springs, and aqueducts. Consequently, when constructing models for terrestrial mobility, our focus remains on inland movement, as these tools are designed for such analysis, while the mobility patterns in the other two environments – rivers and seas – remain uncertain.

The study of river mobility is particularly intriguing as rivers served as vital communication routes in historical times. However, determining the extent of a river's navigability and the types of boats used presents challenges (Villalba 2018). Paradoxically, in terrestrial mobility analyses, rivers are considered constraining factors, despite their clear historical significance in facilitating transportation. For instance, in the Betic province the Guadalquivir River, beyond the capital Corduba, today more than 200 kilometres inland, was navigable during Roman times, exemplifying the presence of fluvial communication. Most significant commercial hubs were indeed located near the coast or along navigable rivers. While we can comprehend the navigability of well-documented rivers such as the Nile (Adams 2007), medium-sized rivers in antiquity, though not necessarily far inland, were navigable at least near their mouths due to favourable geomorphological conditions for boats (Arnaud 2005).

However, it is crucial to note that the current condition of rivers may have undergone significant changes since Roman times, and the precise dynamics of ancient river transport remain elusive. Consequently, a river's navigability could have been limited to only some months of the year, and smaller boats may have been employed. Nonetheless, river transport provided an efficient means of moving heavy loads in an era heavily reliant on human and animal energy resources (Deman 1987). Moreover, we have to consider that in Roman logic, if there was a navigable river, it would be used, but if it were not, it would be a major obstacle. This duality is reflected in the creation of boats to transport goods, and bridges and fords to cross them. In both cases, a simple analysis of land mobility will tend to eliminate it from the calculation or to give it a high restriction, when, with the proposed model, it is possible to model both the transport by river and also to take into account the cost of crossing it.

7.2.1 Conditions

In the case of mobility on water, the main conditioning factors are determined by the type of vessel. The means used can be very different, depending on the river. As long as the boat has a sufficient draught, its displacement is limited by the force of the river current and by the efforts of the boat itself to move. Here we can find oars, sails or even towing from land with ropes. The interesting thing about this type of mobility is that it depends on the orientation of the river, being easier to go with the current than against it.

The main technical problem to make a good historical model of this space is that the conditions of movement are changeable depending on the direction in which we move. As with any mobility analysis, we need to make a friction map, which will hardly reflect reality, given that the river will facilitate mobility when going downstream and vice versa.

The criteria applied to solve these problems can be easily visualized in the case of rivers and streams in historical times. The cost tools mainly consider the slope of the terrain as the main criterion. Therefore, a smooth, straight depression without any obstacles is the area where the cost analysis shows the greatest ease of mobility. However, walking over a river is impossible and crossing certain bodies of water is not advisable (Melles et al. 2011). Even more so if we are calculating the displacement for a loaded transport vehicle. It is necessary to limit mobility through these spaces, as if it were a road in poor condition, indicating to the system that we do not want to travel through that area. To do this, we need a linear entity of streams and rivers and to limit it (Mlekoz 2012), in our opinion, we usually do two things. On the one hand, we can make the passage more difficult, which consists of artificially generating a difficult area, usually by multiplying or adding more slope to the terrain. In this sense, we can multiply, for example, by two the slope value in areas where there is an obstacle that can be bridged such as a stream. Another option would be to multiply a value, for example, by ten in the case of obstacles that are real barriers, such as a lake or a cliff. On the other hand, we have the option of trimming the raster by removing these areas. This is more radical since it prevents the algorithm from considering passing through that area since it does not exist. The problem with using this method is that sometimes it is necessary to move following the path of a river, through its floodplain or crossing it.

Everything described above applies in the case of using the distance cost tool on land and not in itself to modelling water mobility on rivers. In this case, the conditioning factor is a simpler aspect to take into account. Firstly, the capacity to introduce a boat of one type or another into the river, and secondly, the possibility that the speed of movement is modified by external agents or by internal ones.

External agents would depend on the type of boat we are using but suppose at first that we are using small boats in small rivers where there is not a large draught and therefore a large part of the movement is not achieved only by sail navigation but could be by rowing or the current of the river itself. Possibly, we can consider traction by land, be it by donkeys or similar. For that reason, the main conditioning factor is to travel with or against the current. This would modify the possible speed, for example, if there was a rapid or strong current, the possibility of it being navigable is low. So we must always consider ideal conditions since in the case of a rapid current it would be almost impossible to go upstream and therefore also present problems for transporting goods. Therefore, many conditioning factors can be taken into account, not only for the river itself but also in the combination of land and river transport. In our opinion, mobility on the river can be simplified by starting from a fundamental basis, which is the fact of not using this type of element as a limiting factor for movement, but as a possible means of communication, limited in this case by the direction in which the water flows, fundamentally.

Lastly, another aspect to consider is the possibility of traversing a water body. Taking into account the direction of travel, whether upstream or downstream, it is relatively straightforward to incorporate crossings into the analysis. However, the absence of a bridge would significantly increase the cost, potentially rendering it impossible, particularly for a significant portion of the year. On the other hand, if a ford or a designated crossing point, such as a bridge, is available, the situation changes. To identify these specific locations, a dedicated study combining archaeological and topographical analyses becomes necessary (Figure 14).

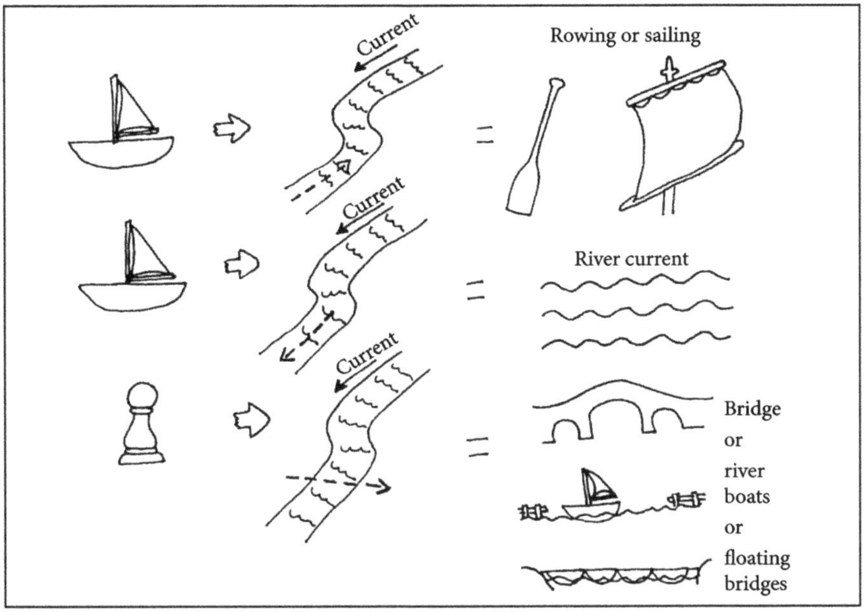

Figure 14 Differences between going upstream, downstream or across a river depending on the direction of movement and the forces or passages involved (source: E. Toscano).

7.2.2 Method

In the case of a river, we can incorporate it into a terrestrial model by adjusting the route of the river itself. This inclusion is essential in our analysis of land mobility since it reflects the historical reality where rivers were used as a means of transportation, rather than being mere constraints.

To calculate this type of constraint, we have developed a methodology (Trapero 2021b) using existing tools for distance cost calculation (Figure 15). This involves employing the horizontal analysis tool, which complements the vertical constraints we have previously examined, such as slopes. For instance, a horizontal factor could be a strong crosswind along a road that affects the movement of vehicles. By determining the angle (measure from the North) of each segment of the river section we wish to analyse, we can establish the distance costs associated with going with, against or across the river.

The horizontal factor is infrequently employed in this particular analytical approach, yet its application proves highly advantageous in this instance. Should the direction of movement align with that of the runoff, it would signify downstream progression. Conversely, a deviation of 180 degrees would denote movement against the flow. Furthermore, angles of 90 degrees or 270 degrees would indicate traversal across the river. It is imperative to acknowledge that the tool will assess angle variations across all conceivable directional shifts.

The significance of this approach lies in the integration of the river into the land mobility analysis, not merely as a constraint but as an additional spatial element. To implement this, we begin by delineating the river line and converting it into a polygon with a buffer to create a wider area. This buffer size is typically 1 or 2 pixels, considering the raster size used in our analysis. Although the river itself may be narrow, we must ensure that it occupies at least the size of one pixel in the raster or add a specific value (e.g., 90 degrees) to represent it.

By combining this modified river layer with the terrain elevation raster, we create a composite raster that incorporates both the terrain height and the fictional cost map. Within this working area, the raster will have a 90-degree slope attributed to the river, distinguishing it from other aspects such as steep slopes. To assign appropriate values, we can use a function table and assign specific values to represent the river's characteristics of interest.

This approach allows us to incorporate the river as a distinct factor with various constraints in terrestrial mobility analysis. Additionally, we need to consider the orientation of each river segment, particularly when considering movement upstream. A simple approach is to apply a cost multiplier when travelling against the flow. The specific values for these multipliers can be derived from previous research and ancient data (de Soto 2009).

Regarding the calculation of the orientation of a line segment, such as a river, we have developed a tool that is discussed in Figure 15. While the process itself is complex, we can summarize the main conditioning factor. The tool, which is available online at Trapero (2021b), involves splitting each line segment and extracting the start and end nodes of each segment. This information is then used to establish a trigonometric relationship that calculates the angle between these points. It is akin to creating a triangle between the two points and determining the angle relative to a specified reference value (0 degrees).

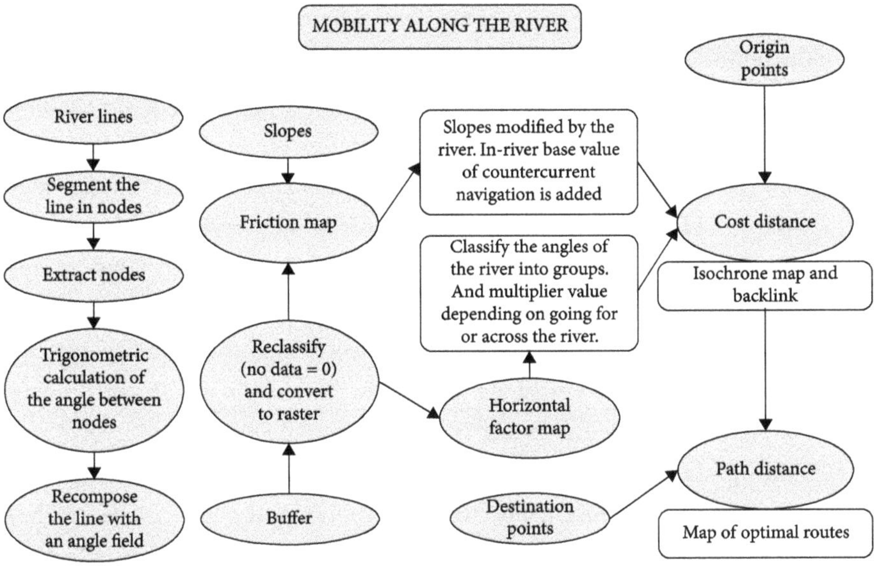

Figure 15 GIS method to calculate river mobility up and downstream (source: author).

Subsequent processes aim to assign this value to each segment, providing a field in the line indicating its orientation in 360 degrees.

7.2.3 Application Examples

This tool has significant potential as the obtained angle values can be incorporated into the distance cost raster as a horizontal value. This process allows us to account for river crossings and consider factors such as downstream or upstream travel, among others.

This approach is much more realistic for understanding the ancient landscape, since, in the case of a river, the possibility of its use as a waterway can be considered. Eventually, this type of approximation can be used to calculate mobility not only for rivers but also for streams. In this sense, misunderstanding the orientation is a good thing, since we could differentiate the cost of crossing a river, going in the direction of the river or going against it, something that could not be done without knowing these directions.

There exist various methodologies for assessing transportation costs along river networks, which hold considerable relevance (Murrieta 2012). In this particular study, the chosen method involves utilizing optimal routes extending from the boundaries of the research area. These routes are determined through the application of a series of algorithms designed to identify the natural passages within the geographical terrain.

Within the confines of this case study, our reference point remains the same scholarly work (Trapero 2021b), as alternative literature on this subject matter is currently limited. This paucity of references may be attributed to the predominant focus of existing studies on terrestrial or maritime transport dynamics, often overlooking the significant role played by rivers in shaping transportation costs. However, it is noteworthy that certain investigations have delved more deeply into this aspect (Fiz and Orengo 2008).

Regarding the application of such exercises, there are currently few prominent examples. Specifically, the cost of transportation on rivers has received little attention in studies due to the aforementioned factors, which make it challenging to model comprehensively. Although there are other perspectives that we could consider similar in their utilization of conditions to comprehend mobility akin to prehistoric times, these perspectives employ various factors such as distance – between sites, inter-visibility, and navigable areas of challenging passage to grasp terrestrial movement in conjunction with maritime mobility (Kealy et al. 2018). Similarly, further instances exist wherein separate methodologies are applied, focusing on major river systems such as the Rhine and the Danube (Werther and Kröger 2017).

The method previously employed could offer a solution to this issue. We recently proposed an application to reconstruct the navigability of Hispania's rivers, which was discussed in greater detail in Section 6.2. The significance of these studies lies in the ability to analyse which rivers are navigable and with what types of vessels. This, in turn, can provide insights into the potential of other routes and shift the paradigm in evaluating the capacity of such principles as a whole.

7.3 Maritime Mobility

In the realm of maritime mobility, analysing mobility is considerably more complex due to a multitude of physical parameters and diverse forces and actions. Furthermore, the advancements in shipping technology over time can greatly influence maritime mobility. When conducting a mobility analysis in this context, it is crucial to consider factors such as wind strength, wind direction, current intensity, swell height, and visibility, particularly in coastal navigation scenarios (Arnaud 2005). Visibility plays a significant role in maintaining sight of the coastline and delineating the navigable area. When seeking to understand maritime mobility in a small, localized area, obtaining precise spatial environmental data might be challenging, necessitating direct engagement with meteorological stations to gather relevant information. However, for broader sea regions, environmental data from international agencies could suffice.

To create an accurate system for analysing maritime mobility, tools like ORBIS (Scheidel and Meeks 2012) can be employed. Although these tools provide a comprehensive and reasonably accurate approach, they may lack the specificity required for our particular case study. Therefore, understanding the functioning of each specific constraint and evaluating their relative importance will enable a deeper understanding of historical data. In this context, the focus should be placed not only on the destination but also on the journey itself.

Generating these models presents three overarching challenges. Firstly, we must determine the appropriate criteria to model these complex relationships, as will be discussed in subsequent points. Secondly, we need to consider the characteristics of the vessels themselves. This requires a solid theoretical understanding of ancient ships (Pomey 2020) and their seaworthiness (Olivella 1994). However, this understanding must be accompanied by studies on the prevailing mindset, as mobility on major trade routes like Ostia (Stöger 2011) differs significantly from smaller-scale, subsistence-based activities such as fishing or other local cultures (Watkins 2015). Thirdly, climatic changes over time can significantly affect maritime navigation, and thus, it is essential to account for these variations. While we assume that the climatic conditions during the Roman climatic optimum were similar to current conditions, the Late Roman period exhibits indications of lower temperatures, potentially influencing the aforementioned navigation factors (Shi et al. 2022).

7.3.1 Conditions

Similar to river mobility, maritime mobility involves a more intricate set of constraints. We have already mentioned some of the factors that need to be considered, which are generally more challenging to model due to limited accessibility. While digital terrain models (DTMs) can provide sufficient information for land-based approximations, creating raster data for the marine environment requires starting from scratch. Current practices in this field involve considering historical navigation conditions, primarily focusing on naval technology, to generate maps of optimal conditions (Leidwanger 2013; Safadi 2016; Gustas and Supernant 2017; Jarriel 2018; Harpster and Chapman 2019). Thankfully, a wealth of satellite data is available today, enabling us to create suitable environmental conditions. For instance, recent experimentation in the Mediterranean scenario explored the seaworthiness of ancient ship technology, taking into account factors such as excessive swell, coastal navigation, and various other constraints (Trapero and Aragón 2022).

As in the previous case, the wind direction has proven to be the most influential parameter in these models, but we have others (Figure 16). Swell is a primary conditioning factor, which can be further classified into three variants:

- Firstly, the strength of the wind itself, typically measured in knots, determines the basic speed of a boat under sail.
- Secondly, the wind direction is crucial, as not all vessels have the capability to take full advantage of all wind directions when navigating in a specific direction.
- Third, wave height plays a significant role, as certain boats have limitations regarding the maximum wave height they can withstand.

Additionally, currents and tides can be valuable considerations and may significantly influence the available options depending on the specific case.

These parameters primarily relate to the ship's internal capabilities, such as its ability to handle friction. However, an essential factor in this regard is the sail's capacity to harness winds from different directions.

Clearly, this question is well suited to considering that environmental changes are constant over time. This is a dangerous assumption, especially if we do not have case studies that allow us to corroborate the data. However, general parameters such as navigation and winds can be extrapolated from the present day (Murray 1987).

Most of these conditions can be measured in the marine environment, provided we have access to environmental data that provide average information on winds, waves, and currents. While these data are generic and require consideration of the current similarity to the historical period under study, they still enable us to perform more or less precise analyses. On a larger scale, international environmental data can be readily utilized. However, for smaller areas, we need to rely on data from meteorological stations and generate our own localized data.

In addition to these internal parameters, we must also account for external constraints:

- The coastline itself, which can undergo significant modifications depending on the historical period of interest.
- Historical information on ports and naval technology, as different types of ships will prioritize certain conditions over others.
- Bathymetry, similar to altimetry, reveals areas where passage is difficult or impossible. Detailed bathymetric information can be instrumental in understanding potential zones of mobility.

With these three factors, we have the foundation for launching an analysis. In coastal navigation, visibility from the coast can be taken into account by

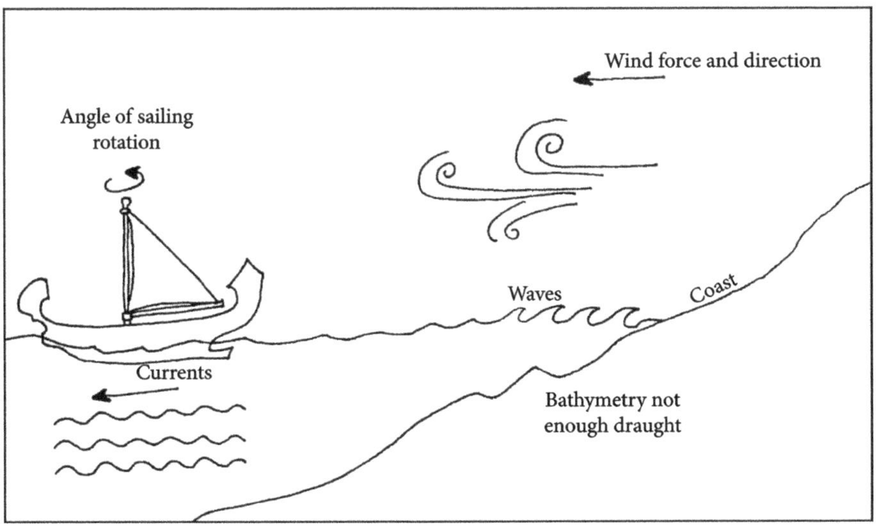

Figure 16 Criteria applied for creating the conditions for sea navigation, considering internal and external factors (source: E. Toscano).

calculating the visible height of the land, accounting for the curvature of the Earth.

7.3.2 Method

In the context of maritime navigation, we encounter a highly complex scenario where vessels are influenced by currents, waves, tides, and winds. Unlike land-based mobility, a ship moves autonomously and must contend with forces that may act against its intended direction. Consequently, determining mobility constraints in the marine environment is considerably more intricate than in terrestrial analysis.

When seeking to access information about navigability, we face the challenge that conditions may have changed since Roman times. It is advisable to calculate averages spanning several years to comprehend the seasonal variations in navigability. By considering different seasons, such as the optimal sailing conditions of summer, multiple analyses can be conducted.

The methodology for conducting such analyses is similar to the approach described earlier. We can generate a friction raster that encompasses the following considerations:

- Water mobility is generally feasible, with two exceptions. Firstly, the shoreline is an obvious constraint. Secondly, areas with very shallow bathymetry would typically be avoided. Both of these areas would be clipped from our friction raster.
- Wind speed serves as the primary conditioning factor, akin to slope in terrestrial analysis. Currents, wave height, and other factors would then be modifiers added to these values. For instance, the cost of traversing an area with strong currents might be twice that of a normal area.

Additionally, it is crucial to measure the wind direction relative to the ship's heading. Just like with rivers, the horizontal factor should be considered. This means that when sailing downwind, the vessel attains maximum speed, gradually decreasing as the angle between the wind and the ship's trajectory increases.

In summary, when conducting maritime mobility analysis, we need to create a friction layer that replaces the slope used in land-based analysis with wind force as the primary factor influencing speed. In the absence of wind, we

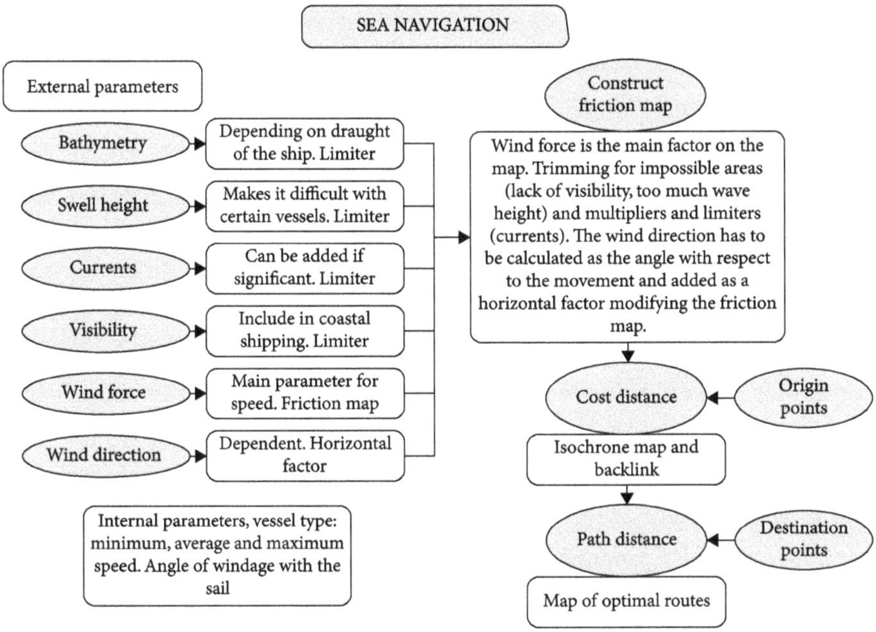

Figure 17 General method to reconstruct sea parameters for modelling maritime mobility (source: author).

assume minimal speed, equivalent to rowing. This friction map is then modified or adjusted to account for additional conditions that impede movement, such as high wave heights, strong currents, or limited visibility. Furthermore, the model requires the inclusion of the horizontal factor, which is the wind direction. When sailing against the wind, the boat's speed will be slower compared to sailing with the wind at its back (Figure 17).

7.3.3 Application Examples

We have already mentioned other case studies throughout this section. In this case, we will briefly present a recent publication where this type of analysis is made (Trapero and Aragón 2022). The results have major implications, being able to model communications in an area such as the western Mediterranean, with a specific type of ship and technology. The difference between different sites is also relevant, as well as the most easily accessible areas. As we are measuring the speed, it is also possible to understand the travel times, revealing which are the main routes and which are secondary (Figure 18).

Other relevant case studies of the applicability of this method can be found in publications such as Safadi (2016). At a theoretical level, it is very similar to the one described above, only the study area changes, in the eastern Mediterranean, and the data source used. Other examples of different chronology could be the case of Indruszewski et al. (2008), about Viking navigation, and Blankshein (2022) for prehistoric navigation,

At a comparative level, these models are particularly difficult to work with, given that the input data are more difficult to process. On a working level, generating a land cost model is much simpler and less time-consuming. It would be necessary to pool these data to generate meaningful models.

In comparison to river navigation, there are other significant examples of maritime navigation (Tartaron 2014; Gambash 2017). Nevertheless, the method employed remains challenging for other researchers to replicate, with the most relevant cases as previously mentioned. It is worth noting that these cases can occur at different scales, which creates a challenge when exchanging methods. When considering navigability in large areas like the Mediterranean, nuances can be lost. Conversely, in specific case studies, we may lack sufficient data precision. Perhaps the best approach would be to establish intermediate-scale models that incorporate both aspects.

In any case, the examples applied still have room for improvement in the sense that a review of capabilities and environmental conditions is needed. The most notable case is the previously mentioned ORBIS website, which estimates these travel times. However, it only does so from centres already defined within the model, i.e. well-known cities. Today, it might be feasible to envision a more complex technology in which the starting and destination points are modelled similarly to Google Maps. Nevertheless, the key to this type of transportation lies in a deep understanding of notable differences in naval technology. This can easily be applicable to other periods, allowing the model to encompass anything from simple, mobility-constrained vessels with limited coastal visibility to modern, large-capacity ships. Thus, if we assume that environmental conditions are similar and, as seen, there is access to significant resources, the produced models can cover a wide diachronic spectrum.

7.4 Transport Costs

A variation of the mobility analyses discussed above is the modelling of transport costs. This approach allows us to gain insights into the historical mindset by considering not only the movement of people but also the transportation of goods. In such cases, we can introduce modifiers or algorithms specific to each mode of mobility, distinct from those used in land-based analysis. For instance, when transporting heavy goods using a wagon, the ability to ascend or descend certain slopes would be greatly limited, thereby imposing constraints on movement (Raepsaet 2022). Including such factors in the analysis can be as simple as eliminating values that fall outside the study range.

However, for this type of approximation, it is necessary to consider not only costs in terms of time and energy but also other concepts, such as the accumulated cost increase. Historically, economic systems have not always operated solely based on supply and demand but have involved different approaches over time and specific needs (De Callataÿ 2014). For example, during Roman times, we can observe the functioning of economic circuits driven by markets, imperial administration, or even euergetism (Duncan-Jones 1974; Remesal 1986; di Salvo 1992; Chic 2006). Therefore, a more realistic model takes into account the specific needs of the era, determining whether a waterway makes the transportation of goods more cost-effective or, conversely, if the priority lies in securing the goods rather than minimizing transportation time. It is worth noting that many products had fixed prices in the market, as exemplified by the edict of Diocletian. This implies that, irrespective of changes in value over time, a significant portion of these commodities would have had low profit margins, making the issue of increasing transport costs more relevant than the time required for transport (Giacchero 1974).

Regarding the cost of transportation, we have already discussed the main determinants, such as the type of vehicle used (land, river, maritime) and the nature of the cargo being transported. It is worth noting that understanding the economy of the ancient world remains a complex issue with various interpretations. Economic activities can be viewed from different perspectives, including economic rationalism, which emphasizes individual interest, cultural factors, religious influence, social dynamics, and political events.

Today, there is no consensus on how the ancient economy functioned, as it can be approached from multiple angles. Some scholars focus on the role of religion and social structures, while others highlight political aspects. Additionally, contemporary research increasingly explores the lives of peasants.

Understanding the ancient economy requires considering three distinct scenarios. The first scenario involves interregional trade, where goods, whether luxury or necessity items are exchanged and endorsed by some form of political power. This can be further categorized into state-managed trade or private trade, depending on the role of political authority.

In the context of Roman history, we see examples of both, such as the state-managed trade facilitated by the *Pax Romana* under Augustus, which followed a period of rampant piracy curbed by Pompey. The type of political environment in place can dictate which trade activities are allowed or suppressed. The objectives of trade can vary, whether driven by exchange, valuing products by their monetary value, or involving barter systems. In some cases, trade may not make sense due to the proximity of neighbouring communities, as exemplified by the exchange of products like chicken eggs with one's neighbour in a reciprocal relationship.

In this case, our focus is not on understanding mobility in terms of time or energy, but rather on examining the changes in the economic cost of transporting a product (Lawton 2004). Fortunately, previous research has addressed this economic increase, which is measured in tonnes per kilometre and represents the cost incurred in transporting goods per unit of weight over a given distance. This concept is commonly used in goods transportation to calculate the cost of each vehicle unit based on the quantity it transports and the area it traverses. Therefore, in certain situations, it may be more economically advantageous to use a motorcycle instead of a truck if the volume of goods being transported is smaller and the distance to be covered is shorter.

By integrating these cost considerations within the same model, we can analyse the costs associated with land, river, and maritime transportation, as discussed earlier. This comprehensive model enables us to identify the most economically efficient communication routes. The routes deemed most optimal in terms of cost may differ from those determined based on other factors, such as transport time.

7.4.1 Conditions

To construct this model, we maintain the conditioning factors discussed in previous cases. However, we need to simplify the model as our focus is on cost rather than strictly calculating time. Neglecting the consideration of time in the transportation of goods makes it challenging to accurately determine their cost. Therefore, certain simplifications are necessary to ensure a successful integration.

One crucial aspect to consider is the type of transportation employed. In this case, we utilize oxen as the primary animals for pulling carts due to their strength, as documented by Otavio (1984). This includes taking into account their stamina and ability to maintain momentum, as noted by Rivas (2019).

While it is possible to analyse the time required for transporting specific products and then apply monetary costs accordingly, this particular version of the model focuses on identifying routes that are optimal in terms of economic cost rather than time efficiency. It is interesting to note that different routes may be deemed optimal when considering time versus economic cost. This distinction holds utmost importance. Therefore, when generating the map, it is recommended to create a fictional representation of the territory that incorporates the economic cost of product transportation.

The cost parameter plays a significant role in route analysis. However, it is generally understood that the economic cost was not considered as crucial in the Roman mindset as it is in contemporary thinking (Baena Sierra, 2016).

In general terms, as depicted in Figure 18, we can classify the cost of distance based on time and economic considerations for both land transport and maritime transport by river or sea (Arnaud 2007). The study of people's transportation poses no significant challenges since time is generally inversely proportional to cost. However, when considering light cargo, the dynamics

Table 4 Transport values in Roman times (adapted based on de Soto 2010)

Transport type	Speed(km/h)	Capacity (t)	Cost (t/km)
Maritime	4,25	92	0,097
Fluvial (downstream)	2,5	5,5	0,33
Fluvial (upstream)	0,6	5,5	0,66
Terrestrial (cart)	1,6	0,090	4,21

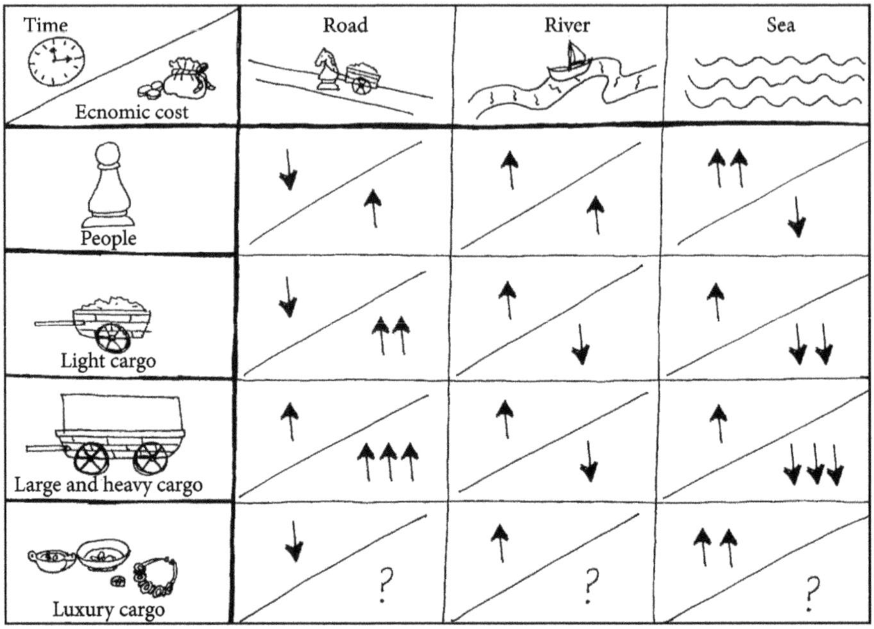

Figure 18 Comparison between the time and economic cost of each type of cargo to be moved and the means used. Arrows symbolize the shortest or longest time with respect to the economic cost of each vehicle compared (source: E. Toscano).

change, as it becomes relatively more expensive on land compared to at sea. This cost difference increases exponentially when dealing with large or heavy cargo, particularly the latter. While river transport serves as an intermediary mode, the comparison between road and sea transport is evident.

It is worth noting that transporting a large or heavy load by ship is essentially the same. In fact, if the ship is not excessively heavy, it needs to be ballasted, whereas a cart would encounter significant difficulties. Additionally, the aspect of luxury goods transportation adds complexity, as we are uncertain about the behaviour of such transactions. It should be considered that for affluent individuals, the economic cost may not carry great significance, and they may be willing to pay extra for the additional cost of the product.

7.4.2 Method

From a terrestrial point of view, the ideal is to limit the territory to areas that are difficult to access, that have steep slopes or that pass through streams, where in this case we are calculating loads. We are talking about the economic cost of moving a person if not a relatively large load; therefore, there will be areas where this type of communication is not feasible. In those cases, we can also include as preferential areas the communication routes that we know exist and therefore generate a raster in the land area that has three basic values. First, a value for where it is very difficult to transit and therefore the cost will be very high. Second, where it would be normal to travel, would be the average cost of land transport. Third preferential areas where we can assume that the cost could be lower.

The cost of the river could be a cost in the middle range of the transport costs because here we are not interested in the time that we take but the increase in cost of transporting the merchandise by river, and finally the same for the maritime area.

By integrating the three criteria discussed earlier, we can create a comprehensive cost map that incorporates maritime, river, and land transportation costs. This map would delineate areas with different cost values, such as optimal land routes, areas where movement is challenging but feasible, and regions with favourable maritime or river transport options (Figure 19). To generate this cost map, we can employ a table-based system with values ranging from one to five, representing the economic cost increase. For example,

Table 5 Example of possible values for each parameter in degrees and their real cost in the incremental value of the product

	Value of friction map	**Correspondence with cost (t/km)**
Slopes between 0 to 5 degrees or on main roads	1	4,21
Slopes between 5 to 15 degrees, difficult to pass	2	6,315*
Slopes more than 15 degrees are hard to cross, or important barriers like big streams	No data	Out of the analysis
The potentially navigable river upstream	3	0,66
The potentially navigable river downstream	4	0,33
Open water areas near the coast or inter-tidal	5	0,097

*Could be modelled as x1.5 more cost, but this is an arbitrary value. In addition, it is possible to estimate the average cost by considering the speed of movement depending on the slope of the terrain and to average the cost more accurately.

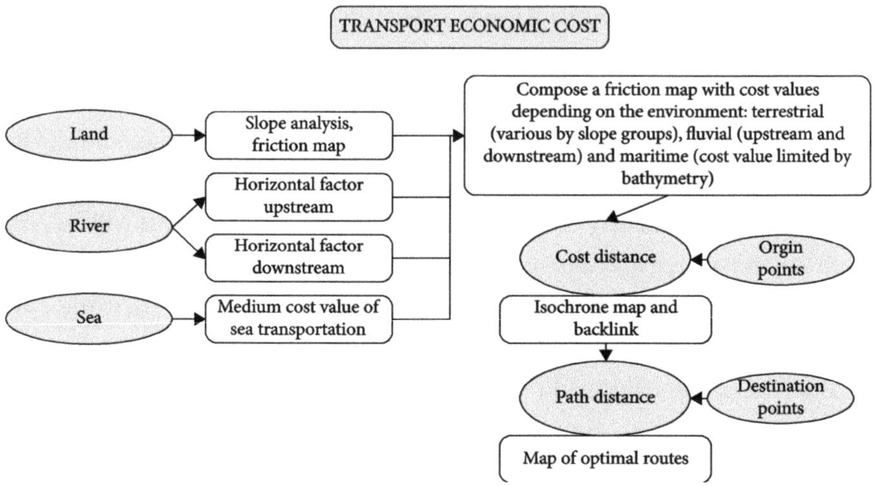

Figure 19 Method for calculating the transport economic cost increase for each type of mobility (source: author).

if a particular region is assigned a value of three for the economic cost increase, it indicates that the transportation of goods in that area incurs a moderate increase in cost compared to other regions. On the other hand, a value of five would signify a significantly higher cost increase, indicating that the transportation of goods in that area is considerably more expensive. By applying this integrated approach and assigning appropriate cost values, we can create a detailed cost map that provides valuable insights into the economic implications of transportation across different terrains.

7.4.3 Application Examples

These approaches, integrating cost analysis into historical models, represent a relatively new development and require further refinement and validation. Nevertheless, we have conducted several case studies to explore their applicability (Trapero 2021a). One notable analysis focused on the transportation of marble, specifically examining the movement of Estremoz marble from the interior of Portugal's ancient Lusitanian province to various regions in Western Europe. This study involved the challenge of transporting heavy loads and determining the most favourable routes. Initial findings indicated that considering the accumulated cost increase, waterways emerged as a more advantageous option compared to land transportation (Trapero et al. 2023a).

Similar examples can be found in the work of Rivas (2019), who modelled the terrestrial transportation of large marble pieces in Greece, considering the cumulative cost of transport. Additionally, studies conducted in the British Isles by Brookes and Huynh (2018) and Orengo and Livarda (2016) proposed somewhat similar models, examining connections primarily by land but also incorporating other means of transportation.

It is worth noting the potential of these tools and their application in generating more accurate models. The integration of cost analysis opens avenues to explore additional parameters related to mentality and symbolic elements. For instance, the chosen route may not necessarily be the shortest but could be influenced by factors such as visiting a nearby shrine for prayer or making offerings. Recognizing and incorporating these elements, treating mobility as a journey rather than a simple point-to-point movement, contributes to the development of more comprehensive models (Carroll and Carroll 2022).

In these applications, there is undoubtedly great potential for a variety of possibilities, especially when considering diverse loads and material types. The main issue, as previously discussed, is the quantitative correlation of values for qualitative products. This can be resolved, as done in previous models, by considering the incremental value of the product. This concept assumes that the value of a product is, for example, 1 and will increase by a certain value due to transportation. In the case of a more expensive product, we only need to know the difference between one and another to propose the variation.

For the application of examples, this presents two main challenges to address. The first is the case of especially significant loads, such as the aforementioned marble, which may require specific vehicles or follow a different logic. Without understanding this aspect, modelling a single model becomes difficult. The second and more significant challenge is that of mixed loads, as it is highly unlikely that this type of transportation would carry only one type of product. This complicates the model but can be achieved as long as this economic aspect is taken into account.

However, and to conclude, it is important to exercise caution when creating overly simplistic economic models, as ancient logic did not always operate in this manner. At times, cost may be disregarded in favour of other considerations, such as being an administrative activity or one driven by philanthropy. In any case, studying the cost increase can help elucidate these matters. For example, questioning the cost of transporting a particular luxury material, the price of which may have increased enormously. In this case, it would allow us to infer which distribution circuit, whether administrative, market-driven, or prestige-oriented, was employed.

8

Visibility Models

One of the valuable functions of historical studies is visibility analysis, which allows us to construct a raster that indicates visible and non-visible areas. Typically, this analysis involves selecting a point as the origin to determine visibility from that point to the surrounding territory. While visibility analysis is a relatively accessible tool that does not require extensive technical knowledge, using it without proper consideration can lead to significant issues.

Most visibility analyses take into account basic parameters such as the height of the observer and the object or person being observed. However, there are additional factors to consider, including the angle of view, Earth curvature, inter-visibility capabilities, and other relevant considerations. Neglecting these factors can result in outcomes that deviate from the desired historical reality. It is important to assess if any modifications to the visibility can be attributed to changes in the landscape itself. Generally, analysing visibility on a small scale should not pose significant challenges, as long as the topographical changes from ancient times to the present are not significant, except for specific cases.

Within a visibility analysis, there are internal parameters that heavily influence and modify the final results. The most critical factors to consider are the height of the observer and the observed object. These heights are typically expressed as vertical distances above the Earth's surface. Determining from which point on the surface a specific area is visible or not visible requires careful consideration of the observer's height and potentially the observed object's height.

The angle of view is another crucial consideration that may require adjustment. Typically, a default window of 90 degrees is chosen, ranging from 45 degrees upward to 45 degrees downward. However, in cases where there are steep slopes, visibility from a high mountain with a significant slope change may not provide a clear view of a nearby large plain due to the angle. In such instances, widening the angle of view is necessary.

It is essential to note that conducting a visibility analysis over a large territory differs from smaller-scale analyses due to refraction and Earth curvature corrections, which are generally handled automatically by most Geographic Information System (GIS) software. Therefore, it is important to evaluate whether the analysis being conducted is meaningful in many cases. Although the human eye is capable of perceiving almost infinite distances, there are limitations imposed by refraction, Earth curvature, and environmental conditions. Moreover, in historical analyses, an object may have been visible, but the observer may not have been able to discern its shape. For example, in archaeology, the focus is often on analysing the visibility of defensive points towards potential assailants or defenders. Thus, this type of analysis may be limited to effective visual ranges, which will be further explored in the study of visibility and defensibility.

Considering these factors, what initially seems like a straightforward analysis can become complex due to multiple variables. Among these variables, the height of the observer, especially if they are a person or positioned on a structure such as a tower, significantly affects the final results. The same applies to the observed point, as it is easier to spot a person on horseback compared to someone on foot. These factors must be taken into account when conducting visibility analyses.

It is common for researchers to choose a central point within the area of interest, such as a Roman city or castle, for visibility analysis. However, this approach can lead to inaccurate visibility results since the central point is unlikely to be the highest point in the area. Additionally, using the highest point as the reference can create complications due to nearby small elevations like vegetation or structures that obstruct the line of sight. To address this issue, it is more realistic to use multiple points for visibility analysis, often referred to as cumulative visibility studies (Wheatley 1995).

Within the examples commonly used in daily life, we have come to understand how, from a specific vantage point, such as a tower or fortress; one can control the rest of the territory. This perspective considers how space is monitored in the event of a potential invasion. However, visibility can be understood from another standpoint.

In practical terms, consider mobility. Imagine a person walking along an ancient road who needs to navigate the terrain. In modern times, we rely on maps and mobile applications to determine our location. However, in ancient times, individuals relied on natural landmarks and geographical markers, such as cardinal points, winds, the position of the sun, and stars, to determine their whereabouts. These reference points were essential for orientation. Roman roads, for instance, had milestones marking the way, allowing travellers to adjust their route as needed.

In this sense, visibility may be underappreciated concerning its significance in mobility. An analysis of visibility from a particular path or point can reveal significant landmarks and reference locations.

Furthermore, visibility plays a role in visual prominence. A specific location's visual prominence can be a component of its significance or prestige. Creating a towering structure that stands out in the landscape can have a profound visual impact. This is evident in the construction of cathedrals, particularly during the Romanesque and Gothic periods, which served as prominent symbols of Christianity, rising above the surrounding terrain. Examining visibility from an optical perspective can be particularly relevant in the context of historical topography.

In this chapter, we focus on the challenges of modelling visibility. While the tool itself is straightforward to use, answering historical questions requires more information and judgment to avoid errors. We will discuss two different approaches depending on the research question: visual control and the identification of sightless areas, and the specific visibility analysis of a site in terms of its defensibility.

8.1 Visual Control and Gaps

To carry out a visibility analysis we have already seen which tools are used, however, we may want to make a more complex model such as inter-visibility and empty areas, like not studying the visibility of a certain site, but to understand the visibility of the whole territory. This model is a little more complex and can be useful for several questions (Zamora 2006). Working with visibility does not indicate the use of visibility calculation tools. Here we are talking about a physical issue, the ability to see or be seen, not in itself the use of the GIS tools of visual watershed or similar. In this sense, it must be understood that there are a series of conditioning factors that greatly modify visibility, depending on what we want to analyse (Brughmans et al. 2014; Ruzickkova et al. 2021).

For instance, visual survey maps can be utilized in archaeological investigations to recognize elements of interest. These maps are particularly valuable in archaeological surveys where detectability maps can be generated based on factors such as soil geology and cultivation practices (Witcher 2008; Casarotto 2018; Casarotto et al. 2021).

As demonstrated, there are numerous applications for such maps. In this study, we specifically focus on assessing the visibility of a large territory to address questions such as: What can be observed from a particular site? What are the inter-visibility relationships between different points? Which strategic spaces are under control? What areas are not visible?

We assume a scenario where visibility plays a crucial role, such as a territory with strategic points that need to be controlled, such as a mountain that could be susceptible to invasion or a coastal region that requires monitoring. In this context, it is important to understand how these locations are interconnected. By examining the key control points within the territory, we can determine if their positioning is influenced by the need for visual control or other factors. Furthermore, the presence of significant gaps in visibility is also noteworthy. Such gaps could indicate a lack of control points in certain areas. There could be various reasons for these gaps, including the absence of settlements in secondary spaces, such as wooded areas, changes in the landscape over time (e.g. drying of a lagoon), or the existence of alternative control points.

Based on the ancient mentality, visual control would be important, especially in the absence of the bird's eye view we have today. In other words, as people orientate themselves by the landmarks of the territory, visibility has a marked character in the defence of a territory, but also in the orientation and control of displacements. In order to control a given property, it would be necessary to see it. The use of the locations promoted by orography has always been relevant and, therefore, knowing what they would be is relatively easy for a model based on given conditions.

8.1.1 Conditions

This type of model is highly valuable for understanding the dynamics of visual capacity. In order to comprehend visibility, it is essential to identify locations that provide visual evidence of this aspect (Van Leusen 1999; Llobera 2007). This includes identifying main settlements, cities, and similar sites, as they are often associated with visual control. However, it is crucial to consider the historical context and other features within the territory, such as towers, fortresses, and castles. Although our focus is on the Roman period, we must acknowledge that sites with visual capacity are not limited to a specific time period. If we have a tower from a later period, it is highly likely that the space had relevance for visual control in earlier and subsequent periods.

Ideally, we should create a comprehensive catalogue of all elements of visual control within the study area, regardless of the time period. This is because these locations likely existed and were reused throughout history. Eventually, we can choose whether to include these points in our analysis. For example, if we assume a tower is only medieval and has no presence in Roman times, but upon conducting the analysis, we observe a specific gap in visual control that could be addressed by that tower, it is highly probable that the tower existed during the Roman period as well, considering the mindset of the *Pax Romana*. In later centuries of stability, many cities moved from higher mountainous areas to lower regions. During that time, the visual aspect may have been less significant. However, these periods occur before and after the conquest and the decline of the Roman Empire.

The second important factor in understanding visual control is not only identifying the locations of these sites but also considering their visual prominence. The nature of these constructions varies depending on whether there were no structures or if there were towers measuring 20 metres in height. It is crucial to incorporate this value into our study (Bernardini 2013). If we lack precise information, we can use a maximum value to determine the potential visibility. The concept is to evaluate whether people in that period aimed to control the entire space and had the means to elevate themselves for enhanced visual control. However, this conditioning factor should be adjusted for environmental factors, as we have previously discussed, as well as the curvature of the Earth. Assuming favourable visibility conditions, there will still be areas that cannot be observed with the same ease due to the refractive index and curvature, which are typically considered in the central analysis. This is particularly relevant when dealing with large study areas (Kormann and Lock 2013). We can also include in our study the key places that must be made visible, i.e. the control, for example, of strategic areas such as a pass or a bridge, certain areas of agricultural and mining use, according to important economic activity or even the population centres themselves.

In this type of analysis, there is little room for caution. We have already indicated that it is easy to push a button and get an analysis, but it will probably be wrong. In this regard, it must be understood that visibility depends on many factors and that the probability of seeing things is not so much a line of 0 and 1 as the tool gives us, but that there are areas of blurred visibility, which, although visible, do not allow us to differentiate elements (Fisher 1994).

Therefore, to consider the distance factor, it is good to start with experimental field data. We can take the following distances in metres as media of possible visibility:

In this sense, it is relatively easy to limit the distance at which a certain element is visible, depending on our needs, similar to the buffer rings mentioned above. A comprehensive overview of these constraints is provided in Figure 20. By considering different ranges of vision, we gain insights into the territorial implications of each range. In general, there are four main categories to consider.

- The first category is the detection area, also referred to as the defence area. This represents the region that can be effectively controlled and monitored. We will delve deeper into this topic in the subsequent subchapter.

Table 6 Range distances based on Fábrega and Parcero (2019)

	Distance range (metres)
First detection of something	2550–2100
Human recognition	1250–975
Basic recognition of the individual	600
Detailed individual recognition	225
Complete identification	60

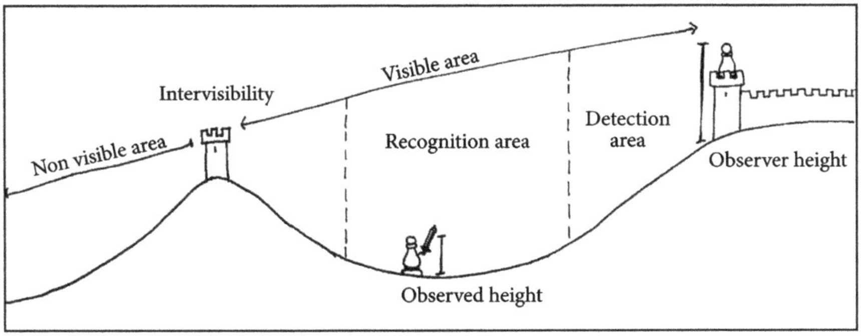

Figure 20 Scheme with the different visible and non-visible areas that can be modelled with the above criteria (source: E. Toscano).

- The second category is the recognition area, where the defensive site has visibility over the territory but lacks direct intervention capabilities.
- The third category encompasses areas with inter-visibility, which serve as warning points between different locations.
- Lastly, we have the non-visible areas, which either are controlled from another vantage point or are not deemed significant for control purposes.

8.1.2 Method

From a methodological standpoint, it is crucial to construct the model not from a single point but from the specific area under analysis. As explained in the visibility section, by considering combined visibility, we can ensure there are no gaps due to potential errors on our part. Additionally, it is important to consider the raster data used for topography. This includes acknowledging whether the model accounts for current conditions or if it is a smooth model with sufficient resolution.

Creating a map that displays visible and non-visible areas is straightforward with the appropriate tools. It is fascinating to explore the relationships between different spaces. For instance, if a tower offers limited visual information to the west and there happens to be an obstacle within that range blocking the view, it suggests a connection between the two.

Furthermore, we should consider other factors that may influence a site's strategic importance, such as access to water sources. This assessment helps determine whether a specific location necessitated visual control in one direction but not in another, or if it could be effectively monitored from an alternative site. It is essential to evaluate whether a site holds historical significance in terms of spatial understanding.

After conducting a comparative visibility analysis from each viewpoint, it is advisable to generate a combined model. This allows for the measurement of inter-visibility between sites (Gillings 2015). The process involves assigning a visibility value of 1 to visible areas and eliminating those with a value of 0, subsequently summing the values across all viewpoints. This approach yields a map wherein the highest number corresponds to areas visible from all the towers. By creating different combinations of spaces, we can assess their historical coherence and identify well-controlled areas versus those that are not. Some spaces may have been left uncontrolled due to chance factors without historical significance, while others may have been deliberately monitored for territorial control.

It is particularly intriguing to consider the spaces that receive a visibility value of 0 and are not visible from any tower. When analysing these areas, it is important to note that small-scale gaps may not necessarily indicate significant findings. However, if these gaps correspond to strategically important locations, it could suggest a lack of visual oversight. In such cases, an inverse analysis can be conducted. This involves selecting the non-visible area, creating a polygon encompassing the general region, converting it into a raster, and generating a polygon with numerous theoretical points surrounding the area. By using the visual tool, we can assess which spaces around the selected area are visible and controlled from these theoretical points.

In summary, this methodology allows for a comprehensive understanding of visibility capacity and its implications (Figure 21). It incorporates both the visibility and strategic aspects of a site, considering the relationships between spaces and identifying areas that require further investigation for their historical significance. Again, if we do a sum, it will give us a lot of visual control points from all the spaces, the points that control more points from those spaces are more likely if there is a tower or visual control place. Therefore, we can go and check if any of the places that have appeared on the map exist. It is a technique that can be used to have a criterion at the time of prospecting.

All of the above methods should be taken into account regarding the limits of vision. In other words, it is not the same to see between control towers, which are designed to be able to see each other and give signals, such as smoke signals, in the event of an attack. However, it must be borne in mind that the first tower must have the ability to identify the attacker. Using the visibility ring data discussed above (Fábrega and Parcero 2019), an average person would be able to see a person at approximately two and a half kilometres but only recognize elements of the same person at approximately half a kilometre. Of course, a lookout's eye could be trained to see further, and an invasion could be seen from further away, because of the group of people or even the dust. What we want to point out here is that the area of vision useful for monitoring space is very limited. To assign a specific number for this is risky, as there are many factors. What we cannot say is that from a tower you can control a mountain pass that is 10 kilometres away, for example.

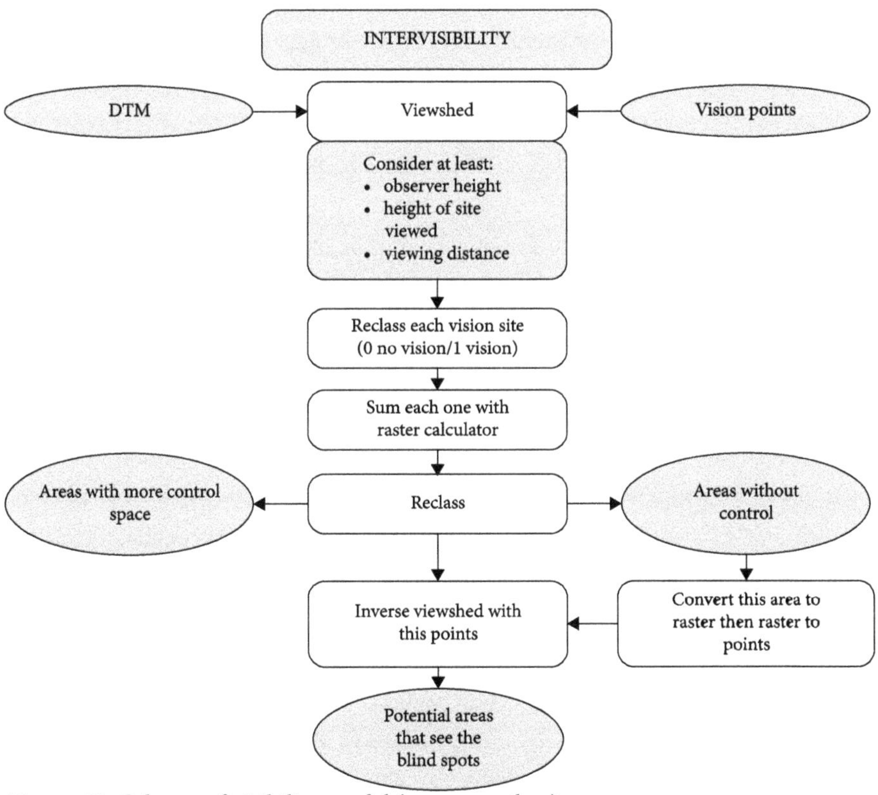

Figure 21 Scheme of visibility model (source: author).

8.1.3 Application Examples

Visibility analysis is a versatile tool with numerous applications, although it is important to acknowledge that some studies may only scratch the surface and lack substantial data for historical modelling. To delve into methodological aspects, the works of Llobera (2007; 2014) offer valuable examples of application. In our own research, we have conducted experiments focusing on understanding inter-visibility between sites, particularly in frontier regions. These investigations aim to calculate the intermediate spaces of inter-visibility (Trapero and Ruiz 2023). In this example, a diachronic study of visibility among notable defensive locations is conducted, necessitating a prior evaluation of these sites as well as their visible potential. This approach enables an understanding of the significance of each site in territorial control, as well as its longevity over time. The examination focuses on a potential border corridor, revealing that both population centres and defensive sites tend to be strategically positioned along it. Conversely, the analysis also considers areas that lack visibility, thereby identifying other vantage points that control the territory but were not originally considered. Similar methodologies are evident in other contexts, such as the Mayan culture (Richards 2017) and prehistoric stone circles (Luke and Ortega 2013), where comparable methodologies are employed.

Nevertheless, there are compelling examples of visibility analysis in other regions that deserve mention. For instance, the work of Turchetto and Salemi (2017) provides an in-depth analysis of the inter-visibility between several significant archaeological sites in Turkey, yielding significant insights.

One intriguing perspective for study is the topographic component of visibility, which remains relatively unchanged over time. This implies that sites with visual prominence are often reoccupied in different historical phases. Consequently, visibility analysis is more commonly approached from a diachronic standpoint rather than focusing solely on a specific moment in time.

Furthermore, while not discussed in detail here, visibility analysis often intersects with mobility studies. For instance, researchers may explore the relationship between a specific communication route and its visibility from other spaces. Conversely, they may investigate how the configuration of landmarks and sites within a territory shapes particular pathways for mobility. Such applications have found utility in diverse fields, such as identifying migration routes in natural areas or studying mobility patterns in prehistoric contexts that share similar considerations.

Nevertheless, we must exercise caution, as we did with mobility, when considering applications of this kind for replication. The apparent ease of the tool may lead to a simple analysis that, once again, lacks significance. As in other cases, the level of detail required will depend on the specific research objectives. In any case, there are significant limitations when conducting such analyses in a specific study area if we do not have highly precise digital models. This is why the most common study scale is typically medium-sized, focusing on indivisibility. Nevertheless, we believe that the potential of these tools is immense, both for studying issues like general visibility and in combination with mobility considerations.

8.2 Site Defence

We have conducted analyses of visibility and its variations in diverse contexts, particularly focusing on large study areas. However, our interest also extends to the analysis of specific areas or individual sites, with a particular emphasis on defence. This aspect has been addressed in the existing literature, although no single method is universally effective. In this section, we will briefly discuss the main considerations and approaches involved in examining the defence of a space.

Many GIS studies in this domain emphasize visibility as a primary factor (Martindale and Supernant 2009; Smith and Cochrane 2011; Zhu et al. 2017), alongside the consideration of mobility (Parcero 2013). The core issue lies in the logic of defending a space. Typically, a wall is constructed to impede or deter hostile individuals, who may attempt to breach or destroy it. In such cases, visual control of the space is crucial to detect the approach of the enemy and effectively defend oneself, capitalizing on the advantages of elevated positions. Consequently, multiple factors can be incorporated into the analysis, as we will explore below.

It is essential to acknowledge that defending a large main space, such as a city, differs significantly from defending smaller and more isolated locations like watchtowers. Furthermore, the topography itself introduces challenges in terms of movement. Consequently, an analysis of defensiveness encompasses more than just visibility, it must also consider the difficulties posed by terrain traversal. Hence, the proposed model integrates components addressing both aspects.

Defensiveness is a relative criterion, contingent upon the level of security provided by a space or structure. Recent research, such as the thesis by Moss (2020), has re-examined some of the problematic aspects and criteria associated with defensiveness. It fundamentally depends on military capabilities, encompassing not only numerical strength but also available resources, and is highly variable based on technological factors. For example, a medieval fortress may prove vulnerable against heavy artillery, not to mention aerial bombardment. Nonetheless, we must start with a basic level of technology to comprehend spaces that possess defensiveness in typical scenarios, impeding individuals with limited resources from gaining access. It is also pertinent to understand which spaces project higher or lower levels of defence. During an assault or siege, attackers will seek out the weakest points among various targets to focus their efforts, while defenders will strive to prevent them from gaining a foothold.

8.2.1 Conditions

The defensiveness of a space depends primarily on several factors. The first factor is the ability to have visual control over the area. Defenders aim to control the space in order to anticipate and repel potential hostile individuals. For instance, there is a zone around a wall where defenders can attack intruders by throwing various objects from the safety and height of the defences. This visual range also acts as a deterrent, preventing unauthorized entry or approach to a specific enclosure. Therefore, elevated spaces such as hills near a wall can become weak points, as the enemy may exploit them to their advantage. In this context, the visible area serves two main purposes. Firstly, it allows for spotting the enemy, preventing attacks, and defending strategic positions. This range can be calculated using the same methods as in the previous model. Secondly, it determines the area from which defenders can effectively attack the enemy, dissuading them from attempting to pass through. Here, a smaller area can be calculated based on the defenders' technical ability to engage the enemy, considering factors such as projectile range based on height and visibility.

Thus, these two values represent the visibility aspect of this type of analysis. However, attackers generally choose to target the weakest point, which is typically the easiest to traverse. Mobility algorithms can be employed to analyse two distinct scenarios. On one hand, we consider the accessibility of the terrain, assessing how easy or difficult it is to navigate. Typically, the fortified positions advantage isolated and elevated landmarks as natural defences. Consequently, the easiest routes to reach these strong points should be well defended. It is possible to calculate the optimal travel route, thus determining the best place to deploy defences. On the other hand, the defensibility of an enclosure can be estimated based on the ease of reaching it from various angles. Different sections of a fortress may have varying levels of accessibility. This variable can be measured in terms of time or energy required.

The combination of these factors determines the defensiveness of a space. If a location is well concealed, it becomes more challenging to breach, but it also poses risks for the defender. Conversely, a space with easy accessibility should be carefully monitored and fortified.

Naturally, numerous conditions can significantly alter our model depending on the technological advancements of the era. For example, improvements in arrow range or defensive measures against such attacks would be significant. Generally, warfare in ancient times followed a similar format over time, so the criteria for defence would be comparable. The main difference lies in the introduction of gunpowder and cannons, rendering certain defences obsolete and necessitating the development of alternative structures, such as star-shaped fortifications. Similarly, contemporary warfare involving aviation presents new challenges. It is important to acknowledge that these technological variations are relevant considerations, and the method employed can be adapted for different historical periods with modifications to the criteria.

Additionally, the concept of defensiveness is of paramount importance. It represents an amalgamation of mobility and visibility, rendering it independent of either factor alone. Conceptually, defensiveness pertains to a space's capacity to obstruct movement and exert control over it, thereby influencing the actions of individuals.

When examining this concept from a diachronic standpoint, we encounter several historical junctures where its significance undergoes substantial shifts. This transformation is intricately linked to the development and utilization of various structures that facilitate the occupation of elevated positions. While there exist prominent examples of such defensive structures, like the fortifications of Heliopolis in Rhodes, it is essential to recognize that this process remained a continuous and pervasive aspect of much of ancient history.

In summary, our analysis using geographic information systems navigates the complexities of multiple landscapes, with a primary focus on rural areas, and explores the concept of defensiveness from a diachronic perspective, reflecting contemporary paradigms in landscape and environmental studies.

Significant changes occurred later on, especially with the introduction of gunpowder. This implied a transformation in fortress design. What used to be a formidable stronghold, requiring weeks, months, or even years for a population to be sieged and starved into submission, could now be relatively easily destroyed at a vulnerable point, causing a breach and enabling capture through force. Similarly, if we understand a certain contemporary scenario, we can see that a particular space is nearly impregnable, as long as it can be defended from the air. Aerial delivery, including airstrikes, drones, and paratroopers, has added a new dimension.

Simplistically, this can be attributed to two primary components, both of which we will explore in this section. Firstly, there is the ability to see the enemy. However, in this context, vision is not solely direct; it also encompasses a shorter range of visibility where the enemy can launch attacks. This constitutes an effective attack range versus defence range, and visibility impacts not only control but also the area where a group of individuals, whether near or far from the fortress in question, can effectively engage in combat.

As depicted in Figure 22, the defence of a site relies on various criteria. In the given example, different sections of the wall are categorized as having low, medium, or high levels of defence based on specific criteria. These criteria typically include the steepness of the terrain, areas with challenging passage, and controlled spaces. For instance, an uncontrolled area represents a location where the enemy can easily move and establish camps without encountering significant obstacles. On the other hand, wall sections that possess defensive

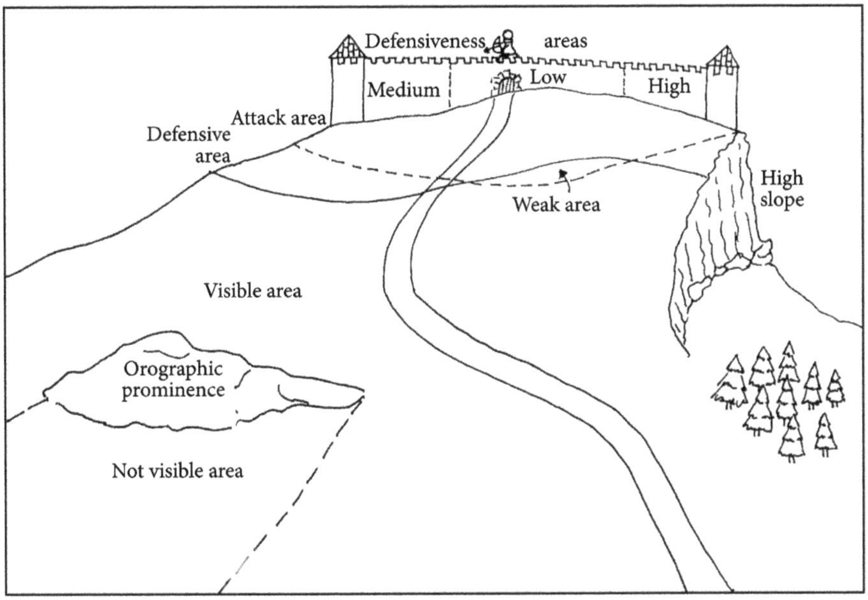

Figure 22 Defensiveness expressed in the terrain factors. Besides the difficulty of movement model with cost tools, we can determine optimum areas for attack and defence purposes (source: E. Toscano).

and visual prominence create secure areas from which attacks can be launched without being easily counterattacked. These areas are where the enemy would contemplate moving. However, it is possible to identify weak areas based on the enemy's line of sight and low defence, creating zones where attackers hold an advantage. These criteria can be modelled, if we have information such as wall dimensions, unaltered topography, and an understanding of the military technology employed during the relevant time. By incorporating these values into the analysis, we can effectively assess the defensibility of different areas within a site.

8.2.2 Method

On this question, we can see different useful methods of action, depending somewhat on the element itself that is being analysed. In this particular case, we will have an enclosure or prominent place where we want to determine its defensive capacity. It should be noted that this type of methodology can be applied to entire territories, even understanding which of the steps leading to a territory would be safer or not. For example, if we have a mountainous area with several mountain passes, we could study the ability to defend them, assigning values. As in other models we have seen, it is not so much the value of the number itself, but the comparative rank between the spaces.

Now, to simplify it, the model described here is based on the assumption that we have a defensive enclosure, such as a wall or natural defences. To make the model, we have to start by having a digital elevation model, with a high resolution. In this aspect, we must not have large anthropic or natural alterations to reconstruct the conditioning factors. In the case of a large area, this is not a problem, but it is necessary to think that the useful area for estimating this defence, in the case of a fortress, can be a few hectares. Therefore, the quality of the data we get will be limited by the spatial resolution. It is highly recommended in this type of work to process LiDAR data, to keep the precise surface, without vegetation. Furthermore, in this way, we will have accurate data.

With this, we must add to the model, employing reclassification of values, the types of soils that we have and the geology. It is interesting to classify those that offer greater difficulty for mobility. Wooded areas can also be included. Of course, this depends to a large extent on the historical knowledge we have, since using a current vegetation layer can be misleading. With all this, it is necessary to simulate with a distance cost the time or better yet, the energy consumed to reach each point or chosen section of the enclosure.

This information must be combined with the visibility analysis, the components of which must be modified for the occasion (Nutsford et al. 2015). In particular, we are not interested in what is seen or what is not, but in the best visual control, so it is advisable to make several rings with distances. The value will depend mainly on the height of the site and the technology.

Both mobility and visibility can be reclassified and added to the field calculator to make a combined map of the areas with less defensiveness. Contrary to other analyses, here we are interested in when the criteria are not met, as these would be the places to attack and defend.

As illustrated in Figure 23, the fundamental approach involves studying visibility, while simultaneously considering a range of additional factors that must be combined with one another. The first factor is general visibility, which identifies areas that are visible or invisible and examines their relationships to understand protected and unprotected regions. The second factor is close visibility, which pertains to controlled areas surrounding the primary defences. This builds upon the previous model of reconnaissance areas and introduces a defence area, where the defender can launch attacks without being vulnerable. The third aspect focuses on observer points, which determine the locations that have a clear line of sight to the walls. By incorporating the distance component and considering military technology, this factor reflects the range of the attacker.

Furthermore, the analysis is complemented by a mobility assessment conducted in two ways. Firstly, the distance cost is evaluated to determine which areas are easier or more difficult to traverse. Secondly, areas with topographical prominence are examined, as they naturally provide a form of defence, such as walls or cliffs. These points are crucial because they cannot be easily accessed by foot soldiers and would require siege equipment, such as ladders, making movement impractical. All of these factors are classified into value groups and weighted using the raster grading method to ascertain whether the necessary criteria are met or not. This comprehensive analysis

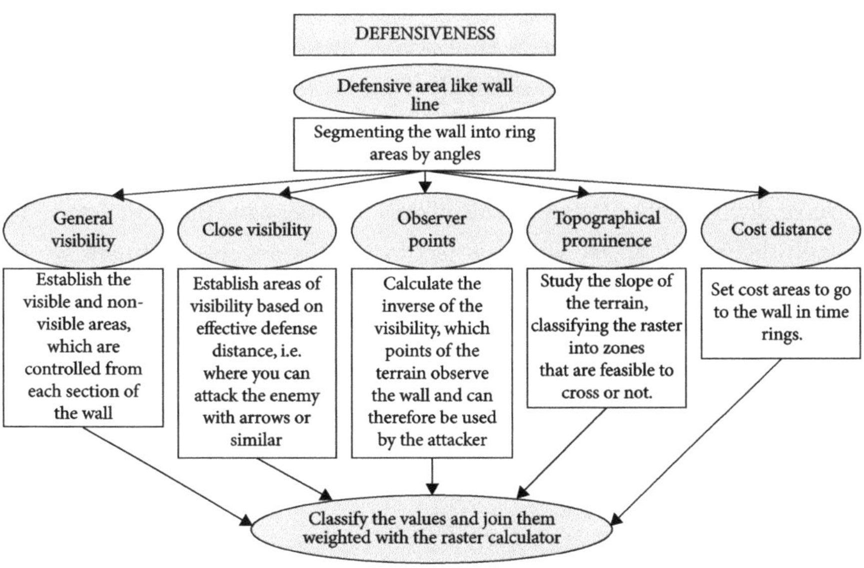

Figure 23 Scheme of criteria used in calculating the defensiveness of a wall line (source: author).

allows for a comprehensive understanding of the defence capabilities and strategic characteristics of the site under study.

8.2.3 Application Examples

The concept of defensiveness is primarily relevant to specific contexts of armed conflicts, limiting its applicability to certain places and historical periods. However, detailed studies can be conducted on specific defensive spaces, such as walls, to gain insights into their strengths and vulnerabilities. Parcero (2013) presents an interesting approach in analysing castles, where the weakest sections of the walls are identified to estimate potential attack points, providing archaeological evidence to support these assessments.

On a larger scale, notable examples can be found in China, particularly in the context of border defences like the Great Wall. These monumental constructions aimed to protect vast territorial areas, necessitating extensive fortifications. Recent studies have attempted to estimate the most and least defensible sections using methodologies similar to the one proposed in this research (Zhu et al. 2017).

However, despite its significant potential, this particular application of the methodology is not extensively developed. We believe it can be applied in two distinct contexts. Firstly, it can be employed to analyse the defence of a fortress itself, providing insights into optimal strategies for attackers and defenders during a siege. Although this application may be anecdotal, it holds great interest for historical narratives, particularly in the field of military history.

Secondly, the methodology can be utilized to understand the most defensible or non-defensible areas within a territory. This involves estimating the likelihood that a military force would choose one route over another based on the difficulty of the terrain or other anthropic conditions. This line of analysis would be particularly valuable in comprehending the reasons behind specific military campaigns. While no current work directly addresses this perspective, it offers an intriguing avenue for future research.

9

Boundaries and Prediction Models

In the final chapter, we will discuss two models that can be applied to the understanding of territories and the prediction of archaeological sites. These models both revolve around the concept of space. The first model aims to establish boundaries for specific places, such as agricultural and livestock sites or cities, by considering the mindset of people in ancient times. By interpreting their criteria, which may still hold relevance today, and examining preserved territorial boundaries or natural features used for demarcation, we can reconstruct these limits to some extent.

The second model focuses on predicting the presence of archaeological sites based on given parameters. Various approaches exist for this purpose, but we will present our own model, which is based on the Roman mentality, as a foundation for applying it to different case studies with changing conditions.

9.1 Settlement Boundaries

In this model, we address the following question: How can we simulate the possible boundaries between cities or rural properties such as *villa*? First, it is crucial to recognize that the Roman conception of space differed significantly from our contemporary understanding. In the modern era, advanced information systems, satellite imagery, and aerial views afford us a vertical perspective of space. We can navigate streets with the aid of mobile phones, orienting ourselves based on digital maps indicating distances and directions. However, in ancient times, as well as just a century ago, individuals did not rely on such vertical perspectives for orientation. Instead, their understanding of space was primarily horizontal, necessitating the use of landmarks within their surroundings.

In ancient societies, individuals navigated and oriented themselves using natural elements such as the position of stars or the sun. Landmarks, such as prominent mountains or rivers, played a crucial role in spatial orientation. Consequently, these elements held immense significance in shaping spatial configurations.

In the context of Roman civilization, we possess a valuable source of information in the form of Roman surveying techniques. Technical literature authored by drafters and engineers tasked with delineating city boundaries and private estates provides insight into the process of spatial demarcation. The approach employed in this process is remarkably consistent across various settings, whether urban or rural, and offers valuable insights into how spatial boundaries were established.

A pertinent example of this approach is encapsulated in a fragment from Siculus Flaccus, which serves as a succinct summary of these fundamental elements:

> *Sic. Fl. De Cond. Agron. 128,17. From the little hill which is so called, to that river, and through that river, to that stream or that way, and through that way to the foot of that mountain, this place is so called; and thence through the top of that hill to the top of that hill, and through the top of that hill, by the watersheds, to the place which is so called, and thence downwards to that place, and thence to the crossroads of that (place), and thence through the (sepulchral) monument of that place to the place from which it (the boundary) first began to be traced.*

Therefore, it should be understood that delimiting boundaries with straight lines, as commonly done nowadays, was not the norm in the past, except in specific cases such as Roman *centuria*. Studies have been conducted on this issue regarding typology (Balil 1964; Chouquer and Favory 2001; Campbell 1996), as well as efforts to determine their layout by examining marks and traces on aerial photographs and LiDAR digital models (Poirier et al. 2013). In general, it can be said that territories were delimited by natural features or, in some cases, by anthropic elements such as divisions, monuments (Compatangelo 1989), ditches, lines of stones, or trees (Dilke 1971). The study of specific places like monuments or sacred spaces is an interesting research direction, as existing models often focus on conditions that do not typically consider these cognitive components. In this regard, there are notable examples where sacred spaces are connected to communication networks (Stek 2014).

Considering these factors, we can simplify the process by incorporating two potential conditioning factors: natural elements that may still be recognizable today and anthropic elements that may not have been preserved. Most territories likely comprise a combination of both. Hence, the challenge lies in the difficulty of modelling this complex interplay. Our approach involves approximating the first part of the process and examining whether the results are consistent with the two outcomes. The first outcome provides an approximate reconstruction that offers historical insights, while the second outcome represents a model that may not align with the logic of the territory. This allows us to identify areas where anthropic boundaries need to be established or plan targeted archaeological surveys more precisely when required.

Consequently, we are going to define specifically the determining factors that must be included in a model to simulate these limits. We start from the premise of knowing mainly the geographical type and we need to specify and know very well where each of these elements is located.

In this case, the main geographical delimiters of a study area can obey three different realities. The first is where we have very clear limiters, for example, a coastline, major river or another series of issues that we can consider as agents within its territory, such as anthropic cases, for example, if we know of a communication route that could also function as a limit. In these cases, the modelling of both a city and smaller properties may be easy, since they would correspond to the precepts of the surveyors.

In this circumstance, it may not even be necessary to make a theoretical model. The need to do so comes above all in the following two cases, especially because they are parameters that we cannot easily represent as a whole. The interest is precisely to be able to represent all the second cases where we do not have such clear, conditioning factors, but where there is some kind of difference in the orography. They are mainly wooded areas with medium or high mountains, multiple streams and rivers. An area that is not flat and that from the Roman point of view would be complicated for *centuriatio*. Considering that the Romans at that time were able to distribute the land in a more or less straightforward way (Chevalier 1961), this would be the third case, which is a space where we do not have important geographical conditions and therefore it is difficult or foreseeable that some of anthropic origin have not been preserved.

It is essential to consider that, from the perspective of demarcating a city; natural features often serve as the primary boundaries. However, a multitude of anthropic actions and political considerations also influence these boundaries, necessitating a comprehensive understanding of how the territory operates. Additionally, to grasp the potential influences at play, we must take into account a crucial detail: the territorial projection of a city, whether it be a colony or a stipendiary city, differs significantly.

As elucidated by Castillo (1996), the total area that a colony, for instance, is capable of controlling will typically be much larger than that of a stipendiary city. Consequently, it is imperative to comprehend this disparity in terms of the potential average distance cost, which may not necessarily align with the actual distance. The model utilized must be tailored to the specific circumstances of each individual case, considering the distinct territorial projections and control capacities of different types of cities.

In any case, this is especially true for delimiting cities, but it can also be used to delimit small spaces, for example, the territory occupied by a specific *villa*, or the total area of a property. In both cases, what we have to achieve are the following conditions.

9.1.1 Conditions

The necessary conditions for this type of model are fundamentally natural, as far as we can currently measure them (Figure 24). The first and most important are the elevations of the terrain and mountainous areas that limit movement and land use. These landmarks are unalterable from antiquity and depending on the type of model should have more or less weight. In the case of studying a city, mountain areas are clear delimiters, but in the case of farms, it is better to use other types such as watersheds.

This is the second constraint. Watersheds are invisible boundaries drawn between the edges of elevations to which the water of a river flows. In this sense, it is also unattainable over time and is the main constraint considered by surveyors. However, watersheds between major rivers can be boundaries that coincide with the previous case, by mountain formations. At another scale level, it may be useful to calculate watershed areas with the corresponding tool, not on the rivers, but on the streams themselves. In this way, we have a line that delimits the runoff water that goes from one stream separated from another.

The third conditioning factor is the river, an area of continuous water that generates barriers to passage. These are very important constraints for agricultural and livestock activities, as we have already mentioned. The exact position of a river or stream may have been modified by meandering or similar over time, so caution must be exercised in their use. It would be relevant to a previous study of how it could be modified over time, to analyse the possibility of huge changes. It must be understood that in ancient times they were areas for public use (Hermon 2014) and that if they are sufficiently deep, they could be fluvial communication areas, as shown in previous models.

A fourth constraint is the streams, discontinuous waters that offer a certain barrier, which may also have undergone minor modifications over time. In this sense, it is more important to understand them as an approximate boundary between farms, as they can be a limiting barrier.

A fifth conditioning factor would be the coastline in maritime environments, which delimits the *villa*, as they are also considered public spaces. In this aspect, it is necessary to stop and study where this limit would be, given that we could make the mistake of delimiting areas that would belong to the sea. This should be taken into account, especially in the Atlantic environment, where the tides are very strong and there may be intertidal spaces only covered by water at times. These areas of preferential use, such as salt marshes, would not be areas to be divided into proprieties, as they would generally be specific enterprises managed or granted by the state (Menanteau 2011).

Additionally, we can consider anthropic factors as a sixth conditioning factor. This category encompasses major known roads, notable features of the landscape like unique geological formations that could serve as landmarks, as well as specific constructions such as sanctuaries or secondary settlements like vici, and other structures along the road such as mansions (Sillières 1990).

Furthermore, we can take into account another conditioning factor: distances themselves. Thiessen or Voronoi analyses are often employed to visually depict the measured distances between the elements under analysis. While useful for illustrating average distances, these methods do not capture the topographic aspect.

In this regard, we propose including a map of isochrones based on mobility calculations as the final conditioning factor. This allows us to visually represent the areas that are closer or further away in the territory. It is essential to understand that not all mentalities are explicitly documented, and certain inherent cognitive factors persist over time. One such factor, which may not be explicitly articulated in historical sources, is the tendency for individuals to prefer locations that are closer rather than farther away. Therefore, it is reasonable to assume that the area of influence and control of a city or villa extends primarily to nearby areas. By constructing an isochrones map based on topography, we can visually discern which places are more accessible. This analysis gains significance when comparing multiple cities or similar entities, as it provides insights into the respective areas of influence. It is important to approach this issue with nuance since individuals may travel for several hours a day to reach their destination or reside not in the main centre under study but in a secondary one. Nonetheless, we believe that incorporating isochrones maps can offer valuable information when integrating these spaces. Areas with better accessibility, closer proximity, are more likely to have stronger

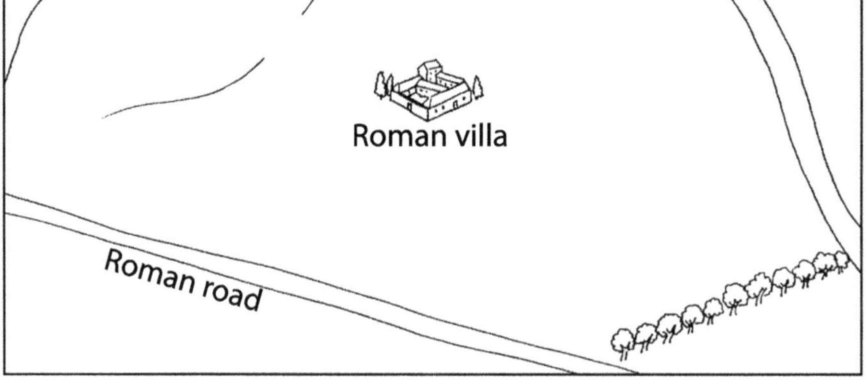

Figure 24 Scheme of the main criteria (natural and anthropic) used in the model (source: E. Toscano).

connections with the organizational centre, while more distant areas may lack other centres due to research gaps or alternative uses, such as *saltus* areas.

9.1.2 Methodology

The model presented here consists of combining all these elements to generate a new layer so that we can easily visualize all the criteria together. That is to say, combining each conditioning factor to see places with a greater possibility of having delimiters in the territory. Therefore, what we are going to do is to choose each one of the elements, reclassify them and add them in a raster layer. Simple, right? Let us see the details.

The first thing we have to do is to transform all the elements that we are going to include in our model since most of them will be linear or polygonal entities for a raster format. After that, the values are reclassified, so that the area where the watershed or the river goes has a positive value, and the rest of the empty territory has a value of 0. The value chosen will depend fundamentally on how we want to make the interrelation. For example, it could simply be a value of 1, so that when they are added together, the higher the number, the more elements of this type coincide at the same time. We do the same with every one of the conditions that we are going to include.

We have to consider which of the conditioning factors mentioned above are relevant for our case study, because if we have a case where we do not have large slopes or where the rivers are not relevant, then we can omit some of them. On the contrary, it is also possible that some elements are more important than others, in which case we may want to give them a higher value. On this question, the ideal is to do several tests, to see which ones fit best with our particular case study. For example, in a flat terrain but with many streams, we can assign a value of 1 to the elevations but a value of 2 to the streams. In this case, we are assigning more weight to one condition than the other in case the first approach would not fit well.

In the case of distance cost, including this value in the model is complicated. The idea is to reclassify it into several time rings, for example, 1 hour, so that we can see several possibilities. In cases where we have very well-contrasted archaeological sites, such as a *villa*, we can look for a reference value that acts as a boundary between the iso-lines of one side and the other. These values would be the maximum topographic distance so that it would take the same time to go from one place to another.

If we consider the mobility between two civic communities, then we need to understand what the edge is in generating that distance cost. This may have to be done independently between each pair of communities. In this way, we will have an iso-line that delimits two communities but at the same time other lines that delimit the next two and the following. This could be a powerful tool to see the vital area of each city concerning the others.

In the case of anthropic delimiters, they can be included in exactly the same way as the others we are talking about, in the case that they are linear or polygonal entities. However, specific key elements of the territory cannot be added as spatial boundaries, because they are punctual, and it is better to leave them for the second phase of analysis.

Once we have every one of these limiters, what we have to do is to assign them a value. As we have already mentioned, it can be a value of 1 all to all of them. In the result where the numbering is higher, the more the constraints coincide. But this is something that is generally not going to happen, because the catchment and the river are not going to overlap. The interesting thing is that it may be that the cost of distance does, and here we have a possible criterion for this delimitation (Figure 25).

In general, the result will be a set of lines without much apparent sense, but in which we have put together on the same map all the spatial conditions. The idea is to generate several combinations of maps showing these points.

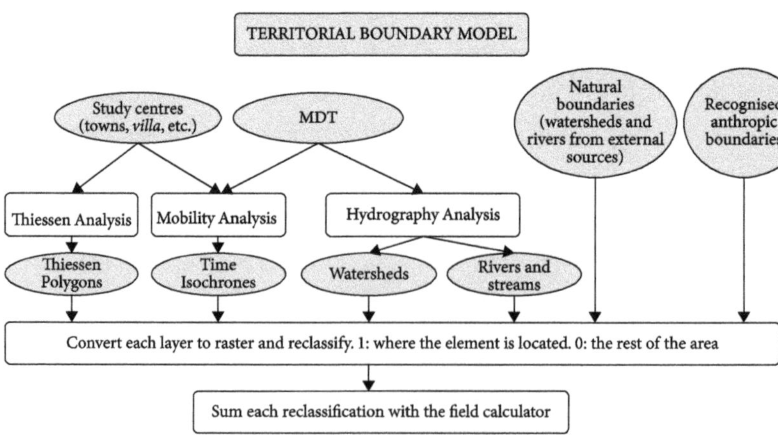

Figure 25 Scheme of territorial boundary model (source: author).

In short, with this method what we are able to do is to represent, cartographically, a series of criteria that appear in surveying and that we can use, not as a definitive map, but as a step to choose which are the best limiters. In this respect, we think it is worth stopping to interpret them, keeping those elements that best fit into the spatial logic of the territory. This method certainly does not represent reality, but it can help us to visualize all the spatial criteria and to make decisions when it comes to establishing boundaries.

9.1.3 Application Examples

In our specific case, we will illustrate the example of the colony of Hasta Regia, which has been the focal point of our thesis (Trapero 2021a). Several spatial analyses have been conducted on this site, and we will revisit it in subsequent examples. Through archaeological surveys, field studies, and analysis of archaeological materials in relevant museums, a series of sites known as *villae* have been delimited. Without delving into the intricate historical details, Hasta Regia was a Roman colony (Martín-Arroyo 2018) likely founded under Caesar or prior to his reign, with a significant territorial extent that is challenging to define. The primary spatial boundary is the marshland at the mouth of the Guadalquivir River, which was navigable in ancient times and referred to as *Lacus Ligustinus* by Avienus (Lagóstena 2016).

The study identified several production centres within the geographically delimited territory. In our case, the coastline itself serves as a key limiting factor. However, we are particularly interested in understanding the vital area of each villa for two main reasons. Firstly, we aim to determine the size of the farms, even if only theoretically. This allows us to assess whether they were of typical, large, or small size and facilitates economic modelling of potential agricultural activities, such as viticulture. Secondly, we seek to analyse how the smaller sites relate to the overall settlement. These smaller rural settlements could serve two different purposes. They could be small properties on the fringes of large estates, either independent or associated with colonists. Alternatively, they may function as dependencies of the villa itself, representing isolated structures.

For these reasons, we have implemented the model described earlier. We consider the coastline as the initial factor, followed by watersheds, streams, cost isochrones, and Thiessen polygons. Neither rivers nor mountains were significant features in the area, as the terrain was predominantly flat, and major rivers were absent.

Another recent case with a similar application can be found on a larger scale in Germany's territorial boundaries. In these cases, the focus is not on delineating individual properties but on entire territories, primarily aligning with the Roman *limes*. The processes and factors considered in these instances are partially similar to those discussed above (Herzog 2022).

9.2 Prediction of Settlements

The last model we can consider is the prediction of archaeological sites. Most models seek to predict certain conditions and are theoretical models, which tell us the optimal places for cultivation or the places where mobility is better, but in general, the aim of them is to understand a theoretical model to present it. There are many examples of the use of this prediction model (Verhagen 2007; Carrer 2013; Caracausia 2018), mostly they tend to be useful for prospection planning and automatic and remote recognition (Van Leusen 2002).

In general terms, a predictive model is born as a borrowed concept from other sciences, where the behaviour of something is simulated based on certain parameters. Although prediction can be applied to anything, in the realm of science, it is about generating a model that represents reality. This is something we have been doing for several chapters, but prediction involves putting it into practice in the real world. In other words, a theoretical model alone is not sufficient; we want to apply it in practice by pinpointing, in our case, the locations of archaeological sites.

Any predictive model used in history faces the significant challenge of being unable to go back in time and verify with certainty whether the model works. Therefore, predictive models are typically used in matters related to the location of archaeological sites, such as individualized units. The location of these sites adheres to a set of criteria that can be measured and recognized in the general context. Thus, predictive models can take two forms: inductive and deductive.

Inductive models begin by drawing conclusions from a large dataset, so that we can almost infer that if all known locations share certain characteristics, then unknown ones will likely share them too. The classic example is that all known crows are black, so one might conclude that all crows are black. Induction has the serious flaw of generalization, but in the case of history, sometimes there is no other option. This is because the model, even if generic, must be tested and sometimes put into practice to arrive at more or less realistic conclusions.

The second case is deduction, in which we start with a reasonably true statement and deduce that individuals behave according to that logic. Again, the general case would be, if all humans are mortal, and I am human, then it follows that I am mortal. Deduction can be combined with induction, so the model is self-reinforcing. This involves creating partial realities and general conclusions so that there are principles that feed off each other. In our case, since we have knowledge of the ancient mentality, we can assume that a group of people will behave according to a certain logic and create the model from there. For example, if we start with the premise that all cities must have a water supply from a spring, we can search for archaeological evidence to see if ancient cities meet this criterion and locate where other unknown sites might exist, thereby enriching the study. This is why we specifically create models to predict archaeological sites to gain a better understanding.

The case of a Roman *villa* can provide a paradigmatic case study. We know that agronomists recommend a predictive model of a particular *villa*, which could be made based on several criteria, like a place where the Romans arrived and set up a colony. In that case, we will have a space in which there could be previous constructions, but possibly divided and assigned in colonial repartitions or similar.

When settlers arrived at a location, they might have encountered existing structures or an empty space, influencing their choice of where to place their *villa*. Whether they adhered to the precepts of agronomists is also uncertain, but the available information considers various criteria or determining factors.

For literary documentation, we have diverse sources providing insights into *villa* placement. Much of this information can be translated into contemporary geographical aspects. To simulate potential locations of archaeological sites, we must consider the spatial logic of priorities, as outlined by Fry et al. (2004) and Espa et al. (2016).

For instance, if we were to model the likely locations of petrol stations today, we could establish a common set of criteria. These might include proximity to major roads, an average distance from neighbouring stations, placement at the entrances and exits of villages, and in areas designated for industrial or commercial activities. While not all petrol stations possess these characteristics, many do. The objective is to understand the feasibility of establishing a series of criteria akin to those outlined below.

9.2.1 Conditions

We can model various types of archaeological sites, depending on the empirical data we have. In particular, it would be interesting to be able to predict almost any but in general, we lack comparative and source data.

Since what we want to predict is whether a particular or general reality holds true, it is ideal to have as much information as possible about the mentality or archaeological reality. Here, the conditions will adapt to the reality of each site. We primarily rely on the ancient mentality, exemplified by Latin agronomists, which, as we have seen in previous chapters, corresponds to a specific reality of an elite and a particular chronology. Therefore, it is highly likely that one or more of the criteria may not be met when generating the ideal conditions.

Fortunately, agronomists provide us with information on crop types, as we have already seen in previous models, but also on where to place the villa. This forms the basis for generating the model, as we describe many constraints that we can include (Figure 26). In general, agronomists agree on a set of conditions (Columella, *Rust.* 1.4.9) that can be exemplified by Varro:

> One should take care to place the house preferably at the foot of a wooded hill, where pastures are extensive, and also exposed to the healthiest winds blowing in the countryside. The one placed facing the equinoxes' rising is the best because it has shade in the summer and sun in the winter. But if you are forced to build near a river, you must be careful not to place it against the current, as it will be much colder in the winter and not healthy in the summer. Also, if there are swampy areas, you should place them in the opposite direction, not only for the reasons mentioned but also because certain tiny animals grow there, invisible to the eyes, and they enter the body through the air via the mouth and nose, causing serious illnesses. 'What can I do,' Fundanius says, 'if I were to inherit a property like this, to minimize the

> harm from the pestilence?' 'I can also respond to this,' Agrius says, 'that you sell it for as many asses as you can, and if you can't, abandon it.' But Escrofa says, 'You must avoid having the house face the direction from which the unhealthiest winds usually blow, and you must not build it deep in a valley but rather in a high place which, as it is well-ventilated, more easily disperses anything that goes against it.' Furthermore, as it is illuminated all day by the sun, it is healthier because also tiny creatures, if any are born nearby and enter, are either scattered by the wind or perish soon due to dryness. Sudden storms and torrents are dangerous for those who have buildings in the lowest and deepest places, as well as bands of robbers who can more easily surprise those who are unprepared. Elevated places are safer for both of these reasons.
>
> Varro, *R.R.* 1.12.1–4

Of course, there are some conditions that cannot be measured or are moderately changeable over time according to agronomists. However, the basis remains the same. In a second phase, we must choose which are the main criteria that most coincide and can be recognized in current geography. This way, we can create a list of criteria to apply to our GIS model. The main criteria are:

- A healthy area, generally in the middle of the hills, away from the cold of winter and the heat of summer. On a small promontory, lest a torrent of water undermines the foundations.
- There should be a perennial water source inside the house or channelled to it. If not, the ability to make a well that is not bitter or brackish. If not, make cisterns for human use and ponds for livestock.
- Firewood and grass in the surrounding area.
- Streams help to moderate the heat and make the landscape more pleasant. If possible, they can be channelled inside the house.
- If the river is very far from the hills, and the elevation of the bank allows it, and the house can be raised on its banks, but be careful to leave the river at the back of the house rather than at the front.
- The house should not face the hostile winds of the region, but the favourable ones. Generally, it should be to the east or south in areas with a healthy atmosphere, and to the north in areas with a harmful atmosphere. The house should be best sited for the equinox, which is the average for winter and summer.

- A house that is splashed by waves is always positive, but there has to be a distance, better to be quite away from the sea, because of the salinity of the water, which deteriorates any type of structure and the possibility of storms.
- It is not advisable to have stagnant water in the vicinity, because of diseases and mosquitoes, and of the humidity, that spoils the property and objects.
- It is not advisable to be next to the main road, because of looting by travellers and of the hospitality that must be given. But not too far either, it is advisable to be close to the road for transport, but not right on the road.

Of the nine criteria outlined, at least seven can be readily modelled in a general sense. However, proximity to timber and pasture areas may not be easily recognizable today due to changes in agricultural practices. Similarly, access to water can pose challenges, as precise hydrological layers indicating springs are often lacking. Without this specific information, and considering the potential for constructing cisterns or transporting water via aqueducts, it becomes difficult to incorporate access to water as a criterion, despite its importance. With more comprehensive knowledge of these resources and

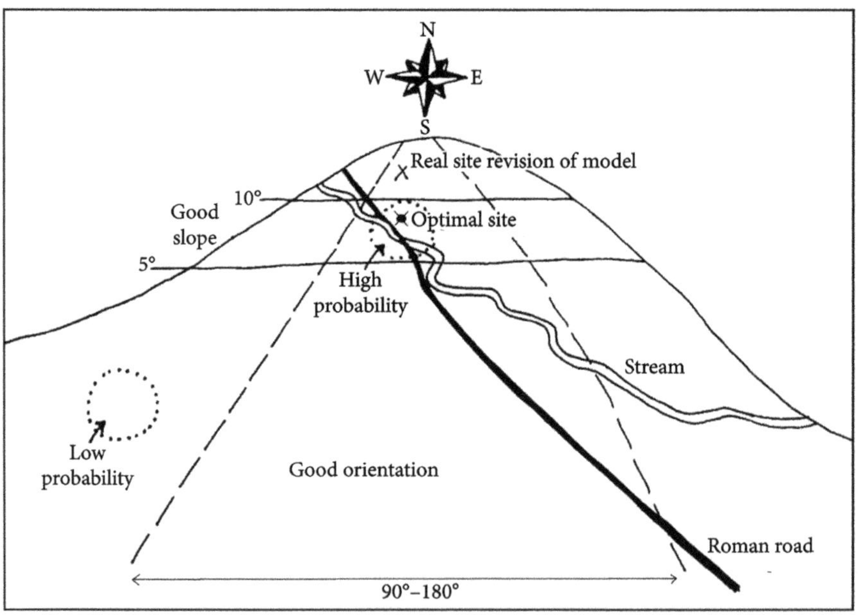

Figure 26 Criteria used in the prediction, with probability areas and the real archaeological site to compare and improve the model (source: E. Toscano).

improved geographic information, it may be possible to include access to water in future studies.

In the context of predictive models, an alternative approach could involve precisely pinpointing the locations of archaeological sites. For instance, points of drinking water collection, such as large springs, have historically been recognizable features. In the case of urban areas requiring significant water supplies, data errors can be minimized due to the presence of a limited number of large springs in the vicinity capable of servicing the city.

Consequently, modelling access to drinking water for rural settlements proves to be quite complex, leading us to focus on the seven criteria mentioned earlier. However, in the case of urban environments, such analyses can be undertaken, as will be elucidated in the application examples section, contrasting with the case of Roman aqueducts.

9.2.2 Method

The methodology employed for evaluating the suitability of different areas involves considering seven conditions, each of which originally could be categorized in binary terms. For example, we may assess whether an area is on a hill, facing east, distant from marshy areas, not too far but also not on the main road, and so on. The method used here, however, is more complex.

The main problem is the assignment of values for this. As you can see in Table 7, we present parameters for each value, which are not the best but need to be refined and specified in multiple models. The great advantage of this predictive model is that it is then applied in reality, and it is checked whether the criteria coincide with the particular case. For example, we have set a distance of 100 metres to consider close or far. This is a logical value, but when applying the model in a specific context, it might be better to use 200 metres because it aligns better with the local reality.

Each condition is represented as a map and classified accordingly. By using the field calculator tool, we can add up these maps to create a composite map where higher values indicate areas that meet all the requirements. The input values are adjusted in a way that resembles an evaluation process.

We have the flexibility to make combinations of these criteria, considering them individually or in relation to each other. We can even assign different weights or importance to each criterion. This can be done using a multiplier to represent the weight as a percentage.

By employing this approach, we can assess the relative importance of different criteria (see Figure 27). There is no strict rule to determine which criterion is more or less important. These predictive models involve considering multiple scenarios to observe any differences between them. If the maps generated from different scenarios are similar, it suggests that one of the criteria may not be significant or representative in that particular territory. It is beneficial to iteratively eliminate or modify the weight of one criterion at a time and observe the similarity of the resulting maps.

It is essential to understand that this process involves assigning weights, which does not imply the absolute importance of one criterion over another. Rather, it reflects the perspective of the time being studied when considering the collective space based on different criteria.

In this sense, it is advisable not to use a binary view of whether a criterion is met or not, but to assess the degree on a scale. For example, the remoteness of a known main Roman road could be classified with a distance cost raster. In this way, it could be divided, for example, into 10 sections, with the first one being the closest not optimal for a *villa*, but the next one could be, and from there the sections decrease in value until the furthest ones. Of course, as in all the models proposed, it is necessary to consider the logic of each specific case. For example, the issue of orientation will depend on the prevailing winds.

In addition, when we measure the importance of any of these criteria, this is the first phase of the analysis, and the second phase is to check it theoretically without any archaeological data to compare with the environmental formation. Then for those areas where it is more probable that certain *villa* appear, we check what is in the territory, and in this way, we can check all the models. In a well-studied area, in order to see if the model works, of all the models that we have produced, there will be one that best matches most of the determining factors, which is the one that probably has a relationship in that territory. If none of the models matches, it is likely that the settlement does not have a distribution with the logic that we see in those sources, so we have to look for other possible influences, let's suppose, for example, the only one that has another settlement pattern or similar.

Table 7 Criteria used in this model with a theoretical assignment of values in reclassification, and weight in the sum

	Type	Range	Value	Weight*1
Slopes of the terrain, optimal will be a medium range	Plain	0 – 5 °	5	
	Medium slope	5 – 10 °	10	20%
	High slope	15 – Max °	0	
The nearness of freshwater also streams for recreational purposes	Water springs	- 100 m	10	
	Streams	- 100 m	10*2	
	Rivers	- 100 m	7	20%
	Others	+ 100 m	0	
Not facing hostile winds	North	315 – 45 °	5	
	East	45 – 135 °	10	
	South	135 – 225 °	5	20%
	West	225 – 315 °	0	
	Plain		0	
The nearness of the coast	Near	- 500 m	10	
	Far	+ 500 m	0	10%
The nearness of marshland and stagnant water	Near	- 500 m	0	
	Far	+ 500 m	10	10%
Proximity to main road	On the road	- 100 m	0	
	Near the road	100 – 1000 m	10	20%
	Far	+ 1000 m	5	

**1* Possible weight that can be used for the analysis. It could be modified by each area's conditions, and also include or eliminate one criterion. **2* The nearness of a stream would always be optimal for the possibility of canalizing water and recreational purposes.

It should be noted that this model allows for a wide range of possibilities, as will be seen in the examples below, but also for corrections and potential improvements. Specifically, it can be useful to use statistical tools such as regression analysis described in the subchapter on statistics and probability. Model correction can be performed in case of not having archaeological data with which to compare or directly for its improvement. In these cases, the methodology to follow is to construct our model with criteria based on ancient mentality from literary sources or archaeological records and then use some tool to assess the statistical consistency of the data. Specifically, decision-making models (Saaty 1994) can be used to indicate whether the parameters provide a consistent model or not. This, although challenging for newcomers, can save time when choosing criteria and their weights, making the method increasingly effective. Of course, we lack comparative studies with similar criteria that would allow us to address these issues.

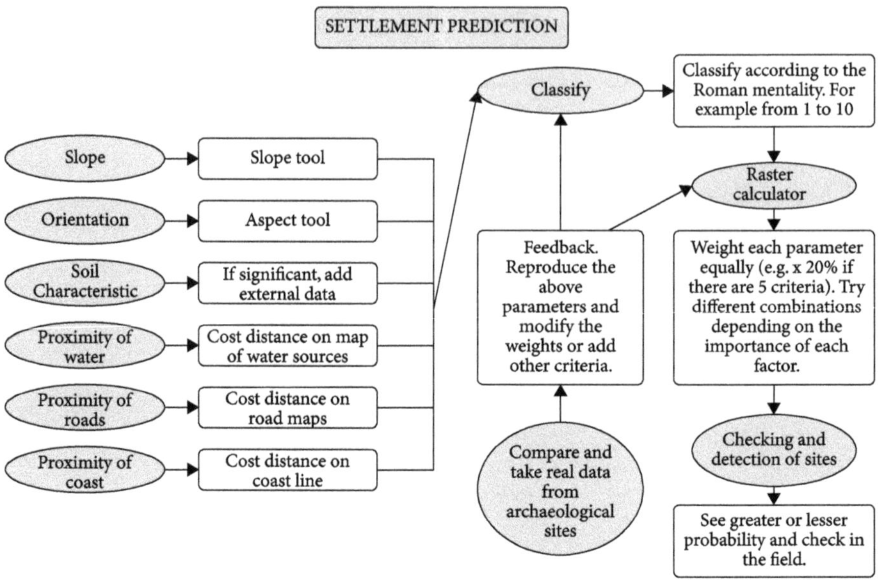

Figure 27 Workflow to predict settlement location, with the feedback in the comparison with real archaeological data (source: author).

9.2.3 Application Examples

There is a wide range of models available as examples for archaeological site prediction, encompassing various contexts and regions. These models extend beyond the Roman world and have been utilized in anthropology as well (Duncan and Beckman 1999; Premo 2004). Noteworthy examples include site prediction in Viking contexts (Stott et al. 2019), Greece (Perakis and Moysiadis 2011), and Israel (Wachtel et al. 2018).

Of the multiple examples that we can find, we see a concentration regarding two great schools or ways of understanding these processes. On the one hand, anthropological, mainly in America, which radiates methods of this science to history and is applied both to Aboriginal communities and to contexts of other cultures. On the other hand, a closer view of the archaeological discovery, especially for the Roman period, starting from an understanding of the literary sources but above all, a great study of the material record.

However, the methodologies employed in these models vary greatly depending on the specific project's needs and objectives. While the outcome may be similar, the processes involved are too diverse to be standardized. For instance, some models utilize criteria based on Roman roads to predict archaeological site locations (Lewis 2021), while others rely on existing data to detect sites directly (Caspari and Crespo 2019).

As an aside, although we are focusing on the detection of archaeological sites such as rural production settlements, the range of application is broader. For example, we could focus on predicting the locations of cities or Roman roads. In the first case, the methodology could be similar to that used for smaller settlements but with different criteria since a city requires various components like defensibility. In the second case, we can determine the location of a stretch of a Roman road and hundreds of kilometres away, the next one. Using cost calculation tools and spatial criteria related to the mentality of people's movement, it would be possible to create a model that predicts where it might go so that we can go to the field and verify it.

We have discussed the issue of water as it relates to creating models, as it is very challenging to include it in the context of water supply for consumption. However, it could be a good candidate for modelling. The method could be similar to what we briefly discussed for roads but with some alterations. If we know the location, to be supplied, the city, and the source of water, the spring, we can, based on ancient mentality, reconstruct the route these spaces would take. A comprehensive systematization of different methodologies in this field can be found in the work of Castiello and Tonini (2021).

In the model described in this chapter, the interesting aspect is that if one of the models proves successful and aligns with the optimal locations identified for *villae*, we can observe other optimal locations as well. In cases where such locations do not align with *villae*, it prompts further investigation to identify the specific settlement type associated with those spaces. This criterion can guide future research and exploration.

Archaeological site prospecting may reveal previously undiscovered sites, some of which may not align typologically with existing records. In such cases, we cannot verify the accuracy of these indicators based on archaeological materials. Additionally, sites may be found in non-optimal locations, raising questions about whether they represent *villae* or other settlement types with different spatial logic. This approach contributes to the prediction and understanding of settlement distribution in general. A recent example of this methodology can be found in the prediction of archaeological sites in Egypt (Manière et al. 2021).

Furthermore, this model can be useful for extrapolation. If we thoroughly analyse a specific case study and understand the settlement patterns around a well-documented city, we can extrapolate that model to areas where information is limited. Even if no sites are currently known in those areas, if the spatial logic remains consistent between the two neighbouring regions, we can anticipate similar settlement dispersion. While there is still much research to be done in this field, defining different models and applying them to various case studies can help develop explanatory models. We will now return to exploring other perspectives.

10

Perspectives and Future Lines of Action

As we have observed, our research has involved the development of GIS models to comprehend specific historical issues and explore ways of modelling the territory and landscape pertaining to those particular inquiries. It is important to acknowledge that while these models are theoretical in nature, many of them are approximations that may not yield definitive results. These models, however, offer a more realistic approach compared to previous methodologies. Although it is unlikely that detractors of this method are among the readers of this book, we anticipate encountering interesting questions if any such individuals are present.

It is evident that these models are theoretical approximations and do not aim to represent reality in a literal sense; rather, they serve as frameworks for understanding the workings of society and economy in antiquity. Historically, most generated historical models have primarily relied on literary and archaeological sources, as well as the researcher's expertise and personal experiences in extensively studying and comprehending those sources.

Nevertheless, this methodology can be challenging for others to sense and comprehend. Consequently, individuals may reproduce this information and perpetuate paradigms that are often continuously replicated. It is plausible that certain sectors may be apprehensive about the direction we are taking, as the application of a particular paradigm can be both correct and incorrect in different cases. In historiography, historical models are frequently replicated from one work to another, but upon revaluation after twenty to thirty years, it becomes apparent that they were founded on the opinions of certain individuals, which may have been well-founded at the time but proven incorrect in light of new data.

The interpretation of historical sources, particularly the understanding of the mentality of individuals, plays a significant role in addressing certain questions. When analysing a text from Julius Caesar's *Gallic Wars*, for instance, it is essential to go beyond mere translation and delve into the original text to grasp the context in which it was produced. If someone claims that something was blue, it is crucial to critically consider the context in which it was written, examining whether there was intentionality behind the statement or if the information provided is accurate. However, it is important to recognize that if we are informed that something was blue, we can certainly question its shade of blue, but we do not possess information suggesting it was green. When someone proposes the latter, they are typically not relying on historical sources.

The theoretical models presented in this book provide researchers with a means to approximate reality and carry out reproducible analyses. These models are based on objective geographical criteria and are mathematically translated into coherent representations. As a result, other researchers can easily replicate and critique the methodologies used. This fosters reproducibility, allowing for the refinement and improvement of models through comparative analysis.

The objective of this book is to provide researchers with a comprehensive understanding of these reproducible models and to encourage their application and refinement in various contexts. Viticulture is one area where these models are already being implemented. Additionally, numerous theoretical models have been developed that can be applied to different historical circumstances and research inquiries. Looking ahead, we anticipate the continued replication and refinement of these models, supported by the establishment of standardized protocols.

The integration of these tools into university and training courses has been widely emphasized. While their direct applicability may vary depending on the research focus and historical period, acquiring spatial representation skills remains valuable. Just as historians of ancient history benefit from understanding writing or palaeography, these tools should be integrated into curricula across disciplines.

Furthermore, there is a need to develop specific tools and platforms tailored to historical research. Existing tools may require adaptation to address the unique requirements of historical analysis, such as the inclusion of historical representations in distance cost analysis. By sharing and integrating these tools within a collaborative program, significant progress can be made in advancing knowledge and understanding.

Establishing an open-access website where researchers can share and access information would be highly beneficial, if feasible. While data sharing may not be a strong tradition in our field, envisioning a scenario where researchers review archaeological data on a unified platform covering a large region is promising. Utilizing online platforms and integrating geographic data directly into the tools would enable researchers interested in these models to conduct analyses within a unified framework and facilitate the reciprocal sharing of information.

The examples discussed above demonstrate the potential of applying GIS to construct models and address historical questions. With the increasing computational capacity and improved accessibility to spatial data, current research holds significant promise for the advancement of this type of analysis.

However, on this issue, the fragile vision of these analyses poses a problem that is linked to their misuse. The variation of a single parameter can give rise to two different results, which gives apparent relativity to the data obtained. This is why there is a historiographical criticism of these points of view. Following Háciguzeller (2012), this type of study can be understood in both representative and non-representational ways. The first implies that the past has an objective and immutable reality that we cannot know directly, so the model to be created must be as faithful as possible to this reality. The second is linked to the fact that the past is not static and that we are continually rediscovering and reconstructing it in the present.

It is this first representative view that we consider the generator of this critical view of history, since the search for the ideal model will always be a constant, and can be detrimental to the value of the different approaches. We need to approach GIS models from a non-representational perspective, as this is how we can observe a constantly evolving investigation and not as an instant solution generated by the tool.

In historical and archaeological research, the utilization of GIS models offers enhanced analytical capabilities, providing an overview of data that would be otherwise challenging to observe due to its vastness. This allows for the analysis of big data by combining historical and spatial information (Verhagen 2018). The generated data can be regarded as a new source of information, necessitating an interpretative process similar to other types of sources, such as literary sources. When examining a text, caution must be exercised in considering translation, chronology, authorial intentionality, and other factors. Therefore, if a literary source is used as an information source, it requires careful handling. It implies that specific statements from a literary source should not be accepted unquestioningly, but rather subjected to a process of revision and critique. Similarly, geographical information modelled with GIS, based on historical considerations, can undergo a similar process.

Moreover, it is crucial to note that when the analysis is not performed by the historian or archaeologist themselves, they should still guide the process. A prime example of this is the tendency to view the river environment solely as a limiting factor rather than as a means of communication, disregarding historical knowledge criteria. This highlights how historical models have sometimes been constrained by the GIS tool, leading to practices that do not align with the conditions and mindset of the past but rather prioritize the convenience of current tool-based analyses. The mobility of land and sea is often assessed based on maximum potential, while the mentality and politics of the historical period are frequently overlooked. For instance, during Roman times, the exportation of products between provinces necessitated the payment of a tax or *portorium*, which was not applicable to all passes and ports (Pasquini 2017). One possible solution to address these challenges would be for historians themselves to undertake this type of approach.

Performing quality, GIS modelling in historical research requires addressing several changes and challenges. Firstly, there is a need for improvements in GIS training within academic curricula. While specific GIS topics are increasingly included, their general application is still lacking. Many processes and algorithms used in modelling require specialized training and a significant learning curve. External training courses are available, but they often focus on teaching the tool itself rather than the historical issues and objectives behind the processes. A more comprehensive integration of GIS training with a focus on historical research is essential.

A second challenge is the digital divide, primarily caused by the high cost of proprietary software. ArcGIS Pro is currently considered the industry standard, but other open-source software as if QGIS is gaining momentum and can potentially achieve similar analysis capabilities. However, the rapid advancement of these tools means that individuals must be trained on specific software versions, which become obsolete within a few years. Additionally, the economic cost of the most powerful software limits access and capabilities for some researchers.

Hardware limitations also pose challenges. While average computers can handle simple GIS analyses, more complex tasks requiring high precision or analysis of large datasets may require more powerful hardware. Technical capabilities and resources become crucial when dealing with advanced spatial analyses.

Access to and generation of spatial information contribute to the digital divide. Different regions or countries may have varying access to spatial data, ranging from high-resolution digital terrain models to satellite-collected data with lower resolution. The availability and sharing of geographic information by public bodies play a role in the development and accessibility of GIS tools. Technological limitations based on available data sources and quality affect the accuracy and effectiveness of the models.

The technological gap is a reality in historical research, although the use of open-source software has gained traction in recent years, aiming to democratize GIS analyses. However, compared to professions like engineering, historical and archaeological research often starts from a lower baseline in terms of GIS training and familiarity with spatial concepts and tools. Bridging this gap and providing equal access to training and resources is crucial.

Thirdly, there is an inherent problem in historical research related to the evaluation process of researchers. The pressure to publish numerous articles in high-impact journals within a short timeframe often prioritizes quantity over quality. In this increasingly commercialized system, publisher and journal indicators may take precedence over the actual quality and contribution of the research. This can lead to the inclusion of GIS technologies as a mere trend without proper understanding or application. It is important to differentiate between using analysis to support research and using numbers to make unfounded assumptions. Analysis should serve as a means to illuminate historical issues, not as a pretext to support predetermined results.

Addressing these challenges requires efforts to improve GIS training within academic curricula, promote the use and development of open-source software, bridge the digital divide in terms of software and hardware capabilities, enhance access to spatial information, and maintain the integrity of research evaluation processes. By ensuring comprehensive and accessible training, historical researchers can harness the full potential of GIS modelling to advance knowledge and understanding in their field.

In addition to the challenges mentioned earlier, there are four key points that need to be reinforced to improve the use of GIS models in historical research.

The first point is to emphasize the value of analysis as a source of information. It is important to develop models that are not definitive but are more faithful to historical reality and the availability of spatial data. This requires asking historical questions and involving professionals in the development and modification of tools. By aligning the models with historical perspectives and objectives, the results can better serve the needs of historical research.

The second point is to continue transferring knowledge about GIS methodologies that can be applied in the historical field. Incorporating techniques such as multi-criteria statistical analysis as a form of decision support can help address historiographical problems that may seem unsolvable. By introducing these methodologies, researchers can explore new approaches and unify criteria to improve historical analyses.

The third point is to prioritize data sharing among researchers and foster collaboration with different research units. Sharing data and methodologies with other institutions and researchers can generate synergies and strengthen the validity and certainty of the approaches. Verification and validation by multiple researchers can enhance the reliability and credibility of the models. Collaborative efforts can also lead to the refinement and improvement of existing methodologies.

Lastly, there is a need to integrate GIS tools into university curricula, similar to how they are integrated into environmental studies. By incorporating GIS training into academic programs, a critical mass of researchers will possess the knowledge and skills to evaluate the importance of different methods and apply them effectively. This integration will also encourage researchers in the field to create new tools specifically designed for historical studies, addressing the unique needs and challenges of historical research using GIS.

By reinforcing these four points, emphasizing analysis as a source of information, transferring knowledge about GIS methodologies, sharing data and collaborating with other researchers, and integrating GIS into university curricula; the future of GIS modelling in historical research can be significantly enhanced. These efforts will promote the development of more accurate and relevant models, foster collaboration and validation, and empower researchers to use GIS tools effectively in their historical investigations.

Bibliography

Adams C. (2007), *Land Transport in Roman Egypt: A Study of Economics and Administration in a Roman Province*, Oxford: Oxford University Press.

Alberti G. (2018), 'TRANSIT: A GIS toolbox for estimating the duration of ancient sail-powered navigation', *Cartography and Geographic Information Science* 45 (6), 510–5281.

Allen K. M. S., S. Green and E. Zubrow (1990), *Interpreting Space: GIS and Archaeology. Applications of Geographic Information Systems*, London: Taylor & Francis.

Amourety A. C. (1986), *Le pain et l'huile dans le Grèce antique*, Besançon: University of Besançon.

Andresen J. (2007), 'Topographic wetness index and prehistoric land use: Reconstructing ancient landscapes and vegetation', in A. Posluschny, K. Lambers and I. Herzog (eds), *Layers of Perception: Proceedings of the 35th International Conference on Computer Applications and Quantitative Methods in Archaeology (CAA),* Bonn: Dr. Rudolf Habelt, 405–10.

Aragón E., P. Argüelles and P. Trapero (2024), 'El río Andarax en época antigua: Análisis de la paleocosta y su territorio', in E. Aragón, P. Argüelles and F. J. Díaz (eds), *Almería y el Mediterráneo entre la tierra y el mar: Estudios comparativos histórico-arqueológicos en el mundo antiguo y medieval*, Vol. 1, 85–110.

Arnaud P. (2005), *Les routes de la navigation antique: Itinéraires en Méditerranée*, Paris: Éditions Errance.

Arnaud P. (2007), 'Diocletian's prices edict: The prices of seaborne transport and the average duration of maritime travel', *Journal of Roman Archaeology* 20, 321–36.

Baek J. and Y. Choi (2017), 'A new algorithm to find raster-based least-cost paths using cut and fill operations', *International Journal of Geographical Information Science* 31 (11), 2234–54.

Baena Sierra J. A. (2016), *Las sanciones relativas al cursus publicus en el Código Teodosiano*, Málaga: Málaga University, Doctoral Thesis.

Baker W. L. (1989), 'A review of models of landscape change', *Landscape Ecology* 2, 111–33.

Balil A. (1964), 'Centuriatio: Observaciones sobre la parcelación y agrimensuras romanas y su reconocimiento', *Revista de Estudios Clásicos* 5, 346–59.

Balla A., G. Pavlogeorgatos, D. Tsiafakis and G. Pavlidis (2014), 'Efficient predictive modelling for archaeolgoical research', *Mediterranean Archaeology and Archaeometry* 14, 119–29.

Banaji J. (2002), *Agrarian Change in Late Antiquity: Gold, Labour and Aristocratic Dominance*, Oxford: Oxford Classical Monographs.

Berglund S. (2001), *GIS in Transport Modelling*, Stockholm: Royal Institute of Technology Department of Infrastructure.

Bernardini W., A. Barnash, M. Kumler and M. Wong (2013), 'Quantifying visual prominence in social landscapes', *Journal of Archaeological Science* 40, 3946–54.

Bernigaud N., A. Bondeau and J. Guiot (2021), 'Understanding the development of viticulture in Roman Gaul during and after the Roma climate optimum: The contribution of spatial analysis and agro-ecosystem modelling', *Journal of Archaeological Science: Reports* 38, 103099.

Bevan A. (2020), 'Spatial point patterns', in M. Gillings, P. Hacigüzeller and G. Lock (eds), *Archaeological Spatial Analysis: A Methodological Guide*, London: Routledge, 60–76.

Bevan A. and J. Conolly (2006), 'Multiscalar approaches to settlement pattern analysis', in G. Lock and B. L. Molyneaux (eds), *Confronting Scale in Archaeology*, Boston: Springer, 217–34.

Bevan A. and M. Lake (2013), *Computational Approaches to Archaeological Spaces*, London: Routledge.

Bintliff J. and K. Sbonias (1999), *Reconstructing Past Population Trends in Mediterranean Europe: The Archaeology of Mediterranean Landscapes*, Oxford: Oxbow Books.

Blankshein S. (2022), '(Sea)ways of perception: An integrated maritime-terrestrial approach to modelling prehistoric seafaring', *Journal of Archaeological Method and Theory* 29, 723–61.

Boak E. and I. Turner (2015), 'Shoreline definition and detection: A review', *Journal of Coastal Research* 21 (4), 12–23.

Böhner J. and A. Oleg (2009), *Land-Surface Parameters Specific to Topo-Climatology*, Volume 33, Chapter 8, Elsevier BV: Amsterdam, 195–226.

Bolstad P. (2003), *GIS Fundamentals: A First Text on Geographic Information Systems*, White Bear Lake: Eider Press.

Brookes S. and H. N. Huynh (2018), 'Transport networks and towns in Roman and early medieval England: An application of PageRank to archaeological questions', *Journal of Archaeological Science: Reports* 17, 477–90.

Brouwer Burg M. (2017), 'It must be right, GIS told me so! Questioning the infallibility of GIS as a methodological tool', *Journal of Archaeological Science* 84, 115–20.

Brughmans T. (2010), 'Connecting the dots: Towards archaeological network analysis', *Oxford Journal of Archaeology* 29 (3), 277–303.

Brughmans T. (2013), 'Thinking through networks: A review of formal network methods in archaeology', *Journal Archaeological Method Theory* 20 (4), 623–62.

Brughmans T. and M. Peeples (2017), 'Trends in archaeological network research: a bibliometric analysis', *Journal of Historical Network Research* 1, 1–24.

Brughmans T., S. Keay and G. Earl (2014), 'Introducing exponential random graph models for visibility networks', *Journal of Archaeological Science* 49, 442–54.

Brun J. P. (2003), *Le vin et l'huile dans la Méditerranée antique: Viticulture, oléiculture et procédés de fabrication*, Paris: Hesperides.

Burford A. (1960), 'Heavy transport in classical antiquity', *The Economic History Review* 13 (1), 1–18.

Campana S. (2018), *Mapping the Archaeological Continuum: Filling 'Empty' Mediterranean Landscapes*, New York: Springer.

Campbell B. (1996), 'Shaping the rural environment: surveyors in Ancient Rome', *The Journal of Roman Studies* 86, 74–99.

Cañizar Palacios J. L. (2016), 'Salubritas e insalubritas: El espacio de los humedales en el imaginario romano', in L. Lagóstena Barrios (ed.), *Lacus autem idem et stagnus, ubi inmensa aqua convenit: Estudios históricos sobre humedales en la Bética* (II), Cádiz, 1–14.

Caracausia S., G. L. F. Berruti, S. Daffaraa, D. Bertèc and F. Rubat Borele (2018), 'Use of a GIS predictive model for the identification of high altitude prehistoric human frequentations: Results of the Sessera valley project (Piedmont, Italy)', *Quaternary International Volume* 490, 10–20.

Carlsen J. (2013), *Land and Labour Studies in Roman Social and Economic History*, Roma: Saggi di storia antica 37.

Carneiro A., P. Trapero Fernández and N. Moreira (2022), 'A exploração do mármore do anticlinal de Estremoz em época romana: discussão dos resultados e perspectivas de futuro', *Mármore 2000 anos de História: Projecto internacional de Historia da Indústria dos Mármores*, Vol. III, 141–55.

Carrer F. (2013), 'An ethnoarchaeological inductive model for predicting archaeological site location: a case-study of pastoral settlement patterns in the Val di Fiemme and Val di Sole (Trentino, Italian Alps)', *Journal of Anthropological Archaeology* 32, 54–62.

Carreras C. (1994), *Una reconstrucción del comercio en cerámicas: la red de transportes en Britannia*, Barcelona: Edicions Servei del Llibre L'Estaquirot.

Carreras C., P. de Soto and E. Muñoz (2019), 'Land transport in mountainous regions in the Roman Empire: Network analysis in the case of the Alps and Pyrenees', *Journal of Archaeological Science: Reports* 25, 280–93.

Carreras Monfort C. (2014), 'Nuevas tendencias y datos sobre la demografía romana en la península ibérica', *BSAA arqueología* LXXX, 53–82.

Carroll F. and E. Carroll (2022), 'Budget travel in the Mediterranean: A methodology for reconstructing ancient journeys through least cost networks', *Journal of Computer Applications in Archaeology* 5 (1), 35–56.

Casarotto A. (2018), *Spatial Patterns in Landscape Archaeology: A GIS Procedure to Study Settlement Organization in Early Roman Colonial Territories*, Leiden: Archaeological Studies Leiden University 43.

Casarotto A. (2022), 'Digitising legacy field survey data: A methodological approach based on student internships', *Digital* 2, 422–43.

Casarotto A., J. García Sánchez, T. D. Stek and J. Pelgrom (2021), 'Surface visibility and the validity of settlement patterns in legacy survey datasets', *The Journal of Fasti Online*, 17.

Cascón Katchadourian J. and J. Alberich Pascual (2021), 'La geoerreferenciación de Cartografía Antigua en los Sistemas de Información Geográficos (SIG): Revisión, Análisis y Estudio comparativo de Softwares de Georreferenciación', *Revista general de información y documentación*, 31.

Caspari G. and P. Crespo (2019), 'Convolutional neural networks for archaeological site detection – finding "princely" tombs', *Journal of Archaeological Science* 110, 104998.

Castiello M. E. and M. Tonini (2021), 'An explorative application of random forest algorithm for archaeological predictive modeling: A Swiss case study', *Journal of Computer Applications in Archaeology* 4 (1), 110–25.

Castillo Pascual M. J. (1996), *Espacio en orden, el modelo gromáatico-romano de ordenación del territorio*. Logroño, Universidad de la Rioja.

Cato, Varro (1934), *On Agriculture*, translated by W. D. Hooper, Harrison Boyd Ash. Loeb Classical Library 283. Cambridge, MA: Harvard University Press.

Cavulli F. and S. Grimaldi (2007), 'To see or not to see: Archaeological data and surface visibility as seen by an AIS (Archaeological Information System) approach', in A. Figueiredo and G. Velho (eds), *Proceedings of the XXXIII Computer Application and Quantitative Methods in Archaeology: The World Is in Your Eyes*, Tomar: CAA Portugal, 21–4 (1).

Cerrillo Cuenca, E. and A. López López (2020), 'Evaluación y perspectivas de uso del LiDAR en la arqueología española', *Boletín del Museo Arqueológico Nacional* 39, 221–38.

Chapman H. (2006), *Landscape Archaeology and GIS*, London: The History Press.

Chase A. S. Z., D. Z. Chase and A. F. Chase (2017), 'Chapter 4 LiDAR for archaeological research and the study of historical landscapes', in N. Masini and F. Soldovieri (eds), *Sensing the Past, Geotechnologies and the Environment*, Berlin: Springer, 16.

Chevallier R. (1961), 'La centuriation et les problèmes de la colonisation romaine', *Ètudes Rurales* 3, 54–80.

Chic García G. (2006), *Economía de prestigio 'versus' economía de mercado*, Sevilla: Padilla Libros Editores & Libreros.

Chouquer G. (2007), 'Els agrimensors romans, cartògrafs de territoris, Mètode', *Revista de difusió de la investigación* 53, 60–7.

Chouquer G. and S. Favory (2001), *L'arpentage romain: Histoire des textes, droit, techniques*, Paris, Editions Errance.

Collar A., F. Coward, T. Brughmans and B. Mills (2015), 'Networks in archaeology: Phenomena, abstraction, representation', *Journal of Archaeological Method and Theory* 22 (1), 1–32.

Collischonn W. and J. V. Pilar (2000), 'A direction dependent least-cost-path algorithm for roads and canals', *International Journal of Geographical Information Science* 14 (4), 397–406.

Columella (1941), *On Agriculture*, Volume I: Books 1–4. Translated by Harrison Boyd Ash. Loeb Classical Library 361. Cambridge, MA: Harvard University Press.

Comet G. (1996), 'Productivité et rendements de céréaliers: de l'histoire à l'archéologie', *Actes des congrès de la Société d'Archéologie Médiévale* 5, 87–91.

Compatangelo R. (1989), 'Le cadastre de Pierre de la Péninsule sallentine: polysémie des formes de limitatio d'un cadastre Antique', *Mefra* 101, 365–402.

Conolly J. and M. Lake (2006), *Geographical Information Systems in Archaeology*. Cambridge: Cambridge University Press.

Cooper A. and C. Green (2016), 'Embracing the complexities of "Big Data" in archaeology: The case of the English Landscape and Identities Project', *Journal of Archaeological Method and Theory* 23 (1), 271–304.

Cooper C. H. V. (2015), 'Spatial localization of closeness and betweenness measures: a self-contradictory but useful form of network analysis', *International Journal of Geographical Information Science* 29 (8), 1293–309.

Costa García, J. M. and R. Casal García (2015), 'Fotografía aérea histórica, satelital moderna y lidar aéreo en algunos recintos militares romanos de Castilla y León', *Portvgalia, Nova Série* 36, 143–58.

Courault C. (2021), 'Organisation du travail autour e la pierre a Cordoue au II eme siècle av. J.C. demographie, agriculture et chaine operatoire', *Romula* 20, 205–50.

De Callataÿ F. (2014), *Quantifying the Greco-Roman Economy and Beyond*, Bari: Edipuglia.

de Herrera A. (1818), *Agricultura General de Gabriel Alonso de Herrera, I*, Madrid: Imprenta real.

de Paz J. M., J. Sanchez and J. F. Visconti (2006), 'Combined use of GIS and environmental indicators for assessment of chemical, physical and biological soil degradation in a Spanish Mediterranean region', *Journal of Environmental Management* 79, 150–62.

de Soto P. (2010), *Anàlisi de la xarxa de comunicacions I del transport a la Catalunya romana: estudis de distribució I movilitat*, Barcelona: Doctoral Thesis.

Déderix S. (2016), 'Traveling across archaeological landscapes: The contribution of hierarchical communication networks', in S. Campana, R. Scopigno, G. Carpentiero and M. Cirillo (eds), *CAA2015. Keep the Revolution Going: Proceedings of the 43rd Annual Conference on Computer Applications and Quantitative Methods in Archaeology*, Oxford: Archaeopress, 555–65.

Déderix S. (2017), 'Communication networks, interactions, and social negotiation in prepalatial south-central Crete', *American Journal of Archaeology* 121 (1), 5–37.

Dell'Unto N. and G. Landeeschi (2022), *Archaeological 3D GIS*, London: Routledge.

Deman A. (1987), 'Réflexions sur la navigation fluviale dans l'antiquité romaine', in T. Hackens and P. Marchetti (eds), *Histoire economique de l'antiquité*, Louvain, 79–106.

Desmet P. and G. Govers (1996), 'A GIS procedure for automatically calculating the USLE LS factor on topographically complex landscape units', *Journal of Soil Water Conservation* 51, 427–33.

Di Salvo L. (1992), *Economia private e pubblici servixi nello Imperio romano: I corpora navicularium*, Messina: Samperi.

Dilke O. A. W. (1971), *The Roman Land Surveyors, an Introduction to the Agrimensores*, Newton Abbot: David and Charles.

Domínguez Petit E. (2004), 'Fuentes literarias para la agricultura cartaginesa: El tratado de Magón', *Habis* 35, 179–92.

Duncan-Jones R. (1974), *The Economy of the Roman Empire: Quantitative Studies*, Cambridge: University Press.

Duncan-Jones R. (1990), *Structure and Scale in the Roman Economy*, Cambridge: Cambridge University Press.

Duncan R. and K. Beckman (1999), 'The application of GIS predictive site location models within Pennsylvania and West Virginia', in K. L. Wescott and R. J. Brandon (eds), *Practical Applications of GIS for Archaeologists*, Colchester: Informa UK Limited, 33–58.

ECMWF (2022), 'European Centre for Medium-Range Weather Forecast. Open Data. Products description', Available online: https://www.ecmwf.int/en/forecasts/datasets/open-data (accessed 20 December 2022).

Ejstrud B. (2003), 'Indicative models in archaeology: Testing the methods', in J. Kunow and J. Müller (eds), *Symposium. The Archaelogy of Landscapes and Geographic Information Systems: Predictive Maps, Settlement Dynamics and Space and Territory in Prehistory. Forschungen zur Archäologie im Land Brandenburg, 8. Archäoprognose Brandenburg I*, Wünsdorf: Brandenburgisches Landesamt für Denkmalpflege und Archäologisches Landesmuseum, 119–34.

El Bouzidi S. (2003), 'La conception de la *villa* rustica chez Caton: Entreprise agricole oú simple ferme rurale?', *Gerión* 21 (1), 167–96.

El-Sheimy N., C. Valeo and A. Habib (2005), 'Digital terrain modelling acquisition, manipulation, and applications', *Proceedings of SPIE* 9323 (2).

Epic (2022), 'Epic Webgis Portugal: Ecological planning, investigation and cartography', Available online: http://epic-webgis-portugal.isa.ulisboa.pt/ (accessed 20 December 2022)

Erdkamp P. (2015), 'Seasonal labour and rural-urban migration in Roman Italy', in L. de Light and L. E. Tacoma (eds), *Migration in the Early Roman Empire*, Leiden: Brill, 33–49.

Espa G., R. Benedetti, A. Meo, U. De Ricci and S. Espa (2006), 'GIS based models and estimation methods for the probability of archaeological site location', *Journal of Cultural Heritage* 7, 147–55.

EU-DEM v1.1. (2022), 'Copernicus, Europe's eyes on Earth: Land monitoring service', Available online: https://land.copernicus.eu/imagery-in-situ/eu-dem/eu-dem-v1.1 (accessed 20 December 2022).

Fábrega Álvarez P. and C. Parcero Oubiña (2007), 'Proposals for an archaeological analysis of pathways and movement', *Archeologia e Calcolatori* 18, 121–40.

Fábrega Álvarez P. and C. Parcero Oubiña (2019), 'Now you see me: An assessment of the visual reconition and control of individuals in archaeological landscapes', *Journal of Archaeological Science* 104, 56–74.

Fernand R., G. Geeven, S. Soetens and V. Klontza-Jaklova (2012), 'Mapping the probability of settlement location for the Malia-Lasithi region (Crete, Greece) during the Minoan Protopalatial Period', in S. J. Kluiving and E. Guttmann-Bond (eds), *Landscape Archaeology between Art and Science*, Amsterdam: Amsterdam University Press, 353–68.

Fisher P. F. (1994), 'Probable and fuzzy models of the viewshed operation', in M. F. Worboys (ed.), *Innovation in GIS, Selected Papers from the First National Conference on GIS Research UK*, Oxfordshire: Taylor & Francis, 161–75.

Fiz J. I. and H. Orengo (2008), 'Simulating communication routes in Mediterranean alluvial plains', in A. Posluschny, K., Lambers and I. Herzog (eds), *Layers of Perception: Proceedings of the 35th International Conference on Computer Applications and Quantitative Methods in Archaeology (CAA)*, Bonn: Dr. Rudolf Habelt, 316–22.

Fonte J., C. Parcero Oubiña and J. M. Costa Garcia (2017), 'A GIS-based analysis of the rationale behind roman roads: The case of the so-called via XVII (NW Iberian Peninsula)', *Mediterranean Archaeology and Archaeometry* 17 (3), 163–89.

Frachetti M. D., C. E. Smith, C. M. Traub and T. Williams (2017), 'Nomadic ecology shaped the highland geography of Asia's Silk Roads', *Nature* 543 (7644), 193–8.

Franconi T. and G. Green (2019), 'Broad and coarse: Modelling demography, subsistence and transportation in Roman England', in P. Verhagen, J. Joyce and

M. R. Groenhuijzen (eds), *Finding the Limits of the Limes: Modelling Demography, Economy and Transport on the Edge of the Roman Empire*, Berlin/Heidelberg: Springer International Publishing, 61–75.

Fry G. L. A., B. Skar, G. Jerpåsen, V. Bakkestuen and L. Erikstad (2004), 'Locating archaeological sites in the landscape: a hierarchical approach based on landscape indicators', *Landscape and Urban Planning* 67, 97–107.

Gambash G. (2017), 'Between mobility and connectivity in the ancient Mediterranean: Coast-skirting travellers in the southern Levant', in E. Lo Casio and L. Tacoma (eds), *The Impact of Mobility and Migration in the Roman Empire. Proceedings of the Twelfth Workshop of the International Network Impact of Empire*, London: Brill, 155–72.

García Armendáriz J. I. (1995), *Agronomía y tradición clásica: Columela en España*, Cádiz: Universidad de Cádiz.

García Moreno L. A. (1979), 'La tecnología rural en España durante la Antiguedad Tardía (SS. V-VII)', *Memorias de Historia Antigua* 3, 217–37.

García Sánchez J. (2013), 'Metodologías de prospección a escala regional y artefactual: la prospección del *Ager Segisamonensis*, comarca Odra-Pisuerga (Burgos)', *Complutum* 24 (1), 9–28.

Giacchero M. (1974), *Edictum Diocletiani et Collegarum de Pretiis rerum venalium*, Genoa: Istituto di storia antica e scienze ausiliarie dell'Universitá di Genova.

Gilgen A., S. Wilkenskjeld, J. Kaplan, T. Kühn and U. Lohmann (2019), 'Effects of land use and anthropogenic aerosol emissions in the Roman Empire', *Climate of the Past* 15 (5), 1885–911.

Gillings M. (2015), 'Mapping invisibility: GIS approaches to the analysis of hiding and seclusion', *Journal of Archaeological Science* 62, 1–14.

Gillings M., P. Hacigüzeller and G. Lock (eds) (2020), *Archaeological Spatial Analysis: A Methodological Guide*, London: Routledge.

Gómez Bellard G. (1996), 'Agricultura fenicio-púnica: Algunos problemas y un caso de estudio', *Complutum Extra* 6 (1), 389–400.

Goodchild H. (2005), 'Modelling Roman demography and urban dependency on central Italy', *TRAC: Proceedings of the Fifteenth Annual Theoretical Roman Archaeology Conference* 15, 42–56.

Goodchild H. (2007), *Modelling Roman Agricultural Production in the Middle Tiber Valley, Central Italy*, Birmingham: University of Birmingham.

Grau I. (2006), *La aplicación de los SIG en arqueología del paisaje*, Alicante: Universidad de Alicante.

Grau I. (2011), 'Movimiento, circulación y caminos en el paisaje digital: La aplicación de los SIG en el estudio arqueológico de los desplazamientos humanos', *Simposio Internacional de Arqueología de Mérida: SIG y Arqueología*, Mérida: Anejos de AESpA, 365–80.

Grau I. (2016), 'Archaeological surveys in areas with high density of artefact: analysis and interpretation proposals', *Quaternary International* 435, 1–10.

Grey C. (2011), *Constructing Communities in the Late Roman Countryside*, Cambridge: Cambridge University Press.

Gustas R. and K. Supernant (2017), 'Least cost path analysis of early maritime movement on the Pacific Northwest Coast', *Journal of Archaeological Science* 78, 40–56.

Guyot A., L. Hubert-Moy and T. Lorho (2018), 'Detecting neolithic burial mounds from LiDAR-derived elevation data using a multi-scale approach and machine learning techniques', *Remote Sensing* 10 (2), 225.

Haciguzeller P. (2012), 'GIS, critique, representation and beyond', *Journal of Social Archaeology* 12 (2), 245–63.

Haisman M. F. and R. F. Goldman (1974), 'Effect of terrain on the energy cost of walking with back loads and handcart loads', *Journal of Applied Physiology* 36, 545–8.

Hall S. (2008), *Topographic Analysis and Predictive Modelling Using Geographic Information Systems*, Clemson: Clemson University, Dissertation.

Harpster M. and H. Chapman (2019), 'Using polygons to model maritime movement in antiquity', *Journal of Archaeological Science*, 111, 10497–516.

Hermon E. (2006), *L'eau comme patrimoine: De la Méditerranée à l'Amérique du Nord*, Québec: Presses de L'universite Laval.

Hermon E. (2009), *Climat et société dans l'Empire romain*, Nápoles: Edipuglia.

Hermon E. (2010), *Riparia dans l'Empire romain: Pour la définition du concept*, Oxford: British Archaeological Reports.

Hermon E. (2014), *Riparia, un patrimonie culturel: la gestion intégrée des bords de l'eau*, Oxford: British Archaeological Reports, International Series.

Herzog I. (2013), 'Theory and practice of cost functions', in F. Contreras, M. Farjasand and J. M. Melero (eds), *CAA2010. Fusion of Cultures: Proceedings of the 38th Annual Conference on Computer Applications and Quantitative Methods in Archaeology*, Oxford: British Archaeological Reports, International Series, 375–82.

Herzog I. (2014), 'A review of case studies in archaeological least-cost analysis', *Archeologia e Calcolatori* 25, 233–9.

Herzog I. (2022), 'Computational approaches towards the understanding of past boundaries: A case study based on archaeological and historical data in a hilly region in Germany', *it – Information Technology* 64 (6), 261–83.

Hesse R. (2010), 'LiDAR-derived Local Relief Models: A new tool for archaeological prospection', *Archaeological Prospection* 17 (2), 67–72.

Hillier A. and A. Knowles (2008), *Placing History How Maps, Spatial Data and GIS Are Changing Historical Scholarship*, Redlands: Esri Press.

Hin S. (2013), *The Demography of Roman Italy. Population Dynamics in an Ancient Conquest Society 201 BCE–14 CE*, Cambridge: Cambridge University Press.

Hollander D. B. (2019), *Agriculture in the Roman Economy*, New York: Routledge.

Howey M. (2011), 'Multiple pathways across past landscapes: Circuit theory as a complementary geospatial method to least cost path for modeling past movement', *Journal of Archaeological Science* 38 (10), 2523–35.

Howey M. C. (2007), 'Using multi-criteria cost surface analysis to explore past regional landscapes: A case study of ritual activity and social interaction in Michigan, AD 1200–1600', *Journal of Archaeological Science* 34 (11), 1830–46.

Indruszewski G. and M. Barton (2008), 'Cost surface DEM modelling of Viking age seafaring in the Baltic Sea', in B. Frescher and A. Dakouri-Hild (eds), *Beyond Illustration: 2D and 3D Digital Technologies as Tools for Discovery in Archaeology*, British Archaeological Reports, 56–64.

Iwahashi J. and R. J. Pike (2007), 'Automated classifications of topography from DEMs by an unsupervised nested-means algorithm and a three-part geometric signature', *Geomorphology* 86, 409–40.

James G., D. Witten, T. Hastie and R. Tibshirani (2021), *An Introduction to Statistical Learning with Applications in R*, New York: Springer.

Janssen V. (2009), 'Understanding coordinate reference systems, datums and transformations', *International Journal of Geoinformatics* 5 (4), 41–53.

Jarriel K. (2018), 'Across the surface of the sea: Maritime interaction in the Cycladic early Bronze Age', *Journal of Mediterranean Archaeology* 31 (1), 52–76.

Kealy S., J. Louys and S. O'Connor (2018), 'Least-cost pathway models indicate northern human dispersal from Sunda to Sahul', *Journal of Human Evolution* 125, 59–70.

Keane R. E., R. A. Loehman, L. M. Holsinger, D. A. Falk, P. Higuera, S. M. Hood and P. F. Hessburg (2018), 'Use of landscape simulation modeling to quantify resilience for ecological applications', *Ecosphere* 9 (9), e02414.

Kenny A. and G. Edwin Fussell (1996), 'L'équipement d'une ferme romaine', *Annales* 21–2, 306–23.

Khatami S. and B. Khazaei (2014), 'Benefits of GIS application in hydrological modeling: A brief summary', *Vatten: tidskrift för vattenvård / Journal of Water Management and Research* 70 (1), 41–9.

Kleinbaum D. G., L. L. Kupper, A. Nizam and K. E. Muller (2007), *Applied Regression Analysis and Other Multivariable Methods*, London: Duxbury Press.

Knappett C. (ed.) (2013), *Network Analysis in Archaeology: New Approaches to Regional Interaction*, Oxford: Oxford University Press.

Kolendo J. (1980), *L'agricoltura nell'italia romana*, Roma: Edit. Riuniti.

Kormann M. and G. Lock (2013), 'Exploring the effects of curvature and refraction on GIS-based visibility studies', in G. Earl, T. Sly, A. Chrysanthi, P. Murrieta-Flores,

C. Papadopoulos, I. Romanowska and D. Wheatley (eds), *Archaeology in the Digital Era: Computer Applications and Quantitative Methods in Archaeology*, Amsterdam: University Press, 428–37.

Lagóstena Barrios L. (2016), 'El Lacus Ligustinus como agente de articulación urbana y conectividad territorial: De las riberas de Hasta Regia a Carissa Aurelia y el acceso a la serranía', *Stvdia Historica* 34, 63–86.

Lake M. and D. Ortega (2013), 'Compute-intensive GIS visibility analysis of the settings of prehistoric stone circles', in *Computational Approaches to Archaeological Spaces*, London: Routledge, 29.

Lawton B. (2004), *Various and Ingenious Machines: Power Generation and Transport*, Vol. 1, Leiden: Brill.

Leidwanger J. (2013), 'Modeling distance with time in ancient Mediterranean seafaring: A GIS application for the interpretation of maritime connectivity', *Journal of Archaeological Science* 40 (8), 3302–8.

Lewis J. (2021), 'Probabilistic modelling for incorporating uncertainty in least cost path results: A postdictive Roman road case study', *Journal of Archaeological Method and Theory* 28 (3), 911–24.

Llobera M. (2007), 'Reconstructing visual landscapes', *World Archaeology* 39, 51–69.

Llobera M. and T. Sluckin (2007), 'Zigzagging: theoretical insights on climbing strategies', *Journal of Theoretical Biology* 249 (2), 206–17.

Llobera M. (2018), 'Visibility modelling', in S. L. López Varela (ed.), *The Encyclopedia of Archaeological Sciences*, New Jersey: John Wiley & Sons, https://doi.org/10.1002/9781119188230.saseas0598

Llobera M., P. Fábrega Álvarez and C. Parcero Oubiña (2019), 'Order in movement: a GIS approach to accessibility', *Journal of Archaeological Science* 38, 843–51.

Lo Cascio E. (1994), 'The size of the Roman population: Beloch and the meaning of the Augustan census figures', *Journal of Roman Studies* 84, 23–40.

Lo Cascio E. (2009), *Crescita e declino: studi di storia dell'economia romana*, Rome: L'Erma di Bretschneider.

López Castro J. L. (2012), 'La influencia fenicia y cartaginesa en la organización del territorio hispano', in J. Santos Yanguas and G. Cruz Andreotti (eds), *Romanización, fronteras y etnias en la Roma antigua: el caso hispano*, Vitoria: Pais Vasco University, 113–42.

Lozic E. and B. Stular (2021), 'Documentation of archaeology specific workflow for airborne LiDAR data processing', *Geosciences* 11, 26.

Lufafa A., M. M. Tenywa, M. Isabirye, M. J. G. Majaliwa and P. L. Woomer (2003), 'Prediction of soil erosion in a Lake Victoria basin catchment using a GIS-based Universal Soil Loss model', *Agricultural Systems* 76, 883–94.

Magna (2022), 'Cartografía del IGME, Serie de cartografía geológica y mapas regionales. Serie Magna y Geode', Available online: https://info.igme.es/

cartografiadigital/geologica/default.aspx?language=es (accessed 20 December 2022).

Manière L., M. Crépy and B. Redon (2021), 'Building a model to reconstruct the Hellenistic and Roman road networks of the eastern desert of Egypt: a semi-empirical approach based on modern travelers' itineraries', *Journal of Computer Applications in Archaeology* 4 (1), 20–46.

Marlière E. (2001), 'Le tonneau en Gaule romaine', *Gallia* 58, 181–201.

Márquez J., I. Vallejo and J. I. Álvarez (2015), 'Estimación del tiempo de demora en rutas pedestres: comparación de algoritmos', *GeoFocus* 15, 47–74.

Martín R. (1971), *Recherches sur les agronomes latins et leurs conceptions économiques et sociales*, Paris: Societé d'édition les belles lettres.

Martín-Arroyo Sánchez D. (2016), 'Modelización de la ratio *riparia/vinea*: el emparrado romano entre *Hasta Regia* y *Gades*', in L. Lagóstena Barrios (ed.), *Lacus autem ídem et stagnus, ubi inmensa aqua convenit: Estudios históricos sobre humedales en la Bética II*, Cádiz, 105–25.

Martín-Arroyo Sánchez D. J. (2018), *Colonización romana y territorio en Hispania: el caso de Hasta Regia*, Universidad de Barcelona, Colección Instrumenta.

Martín-Arroyo Sánchez D. and J. Remesal Rodríguez (2018), 'Modelado GIS de la agricultura romana en un marco colonial teórico: Vegetación de ribera y viticultura en Hasta Regia', *SPAL – Revista de Prehistoria y Arqueología* 27 (1), 211–35.

Martín-Arroyo Sánchez D. and M. Castro García (2019), 'GIS-based modelling for the Riparia/Vinea Ratio: From Portus Gaditanus to Nabrissa Veneria', in J. Remesal Rodríguez, V. Revilla Calvo, D. J. Martín-Arroyo Sánchez and A. Martín i Oliveras (eds), *Paisajes productivos y redes comerciales en el imperio romano*, Barcelona: Universidad de Barcelona, Serie Instrumenta 65, 73–88.

Martín-Arroyo Sánchez D. and Trapero Fernández P. (2015), 'La explotación romana de la vegetación riparia y el análisis espacial de la ocupación rural', in L. Lagóstena Barrios (ed.), *Qui lacus aquae stagna paludes sunt*, Cádiz: Seminario Agustín de Horozco, 243–60.

Martindale A. and K. Supernant (2009), 'Quantifying the defensiveness of defended sites on the northwest coast of North America', *Journal of Anthropological Archaeology* 28, 191–204.

Mayoral Herrera V., C. Parcero Oubiña and P. Fábrega Álvarez (2016), *Archaeology and Geomatics: Harvesting the Benefits of 10 Years of Training in the Iberian Peninsula (2006–15)*, Leiden: Sidestone Press.

Meijer F. and O. Van Nijf (1992), *Trade, Transport and Society in the Ancient World: A Sourcebook*, London: Routledge.

Melles S. J., N. E. Jones, B. Schmidt and B. Rayfield (2011), 'A least-cost path approach to stream delineation using lakes as patches and a digital elevation model as the cost surface', *Procedia Environmental Sciences* 7, 240–5.

Menanteau L. (2011), 'Las Salinas', in J. Rubiales Torrejón (ed.), *El río Guadalquivir, del mar a la marisma*, Vol. 2, Sánlucar de Barrameda, Andalucía: Junta de Andalucía, 376–7

Milheira R. G., J. G. Souza and J. Iriarte (2019), 'Water, movement and landscape ordering: A GIS-based analysis for understanding the mobility system of late Holocene mound-builders in southern Brazil', *Journal of Archaeological Science* 111, 105014–26.

Mínguez García, M.C. and E. Capdevila Montes (2016), *Manual de tecnologías de la información geográfica aplicadas a la Arqueología*, Madrid: Comunidad de Madrid.

Miraglia M. (2016), 'Aplicaciones de la cartografía histórica y las tecnologías de la información geográfica en la historia ambiental', *Revista de Histórica Regional* 24 (1), 24–41.

Mishra N., R. Schreiber, I. Stanton and R. E. Tarjan (2007), 'Clustering social networks', in *Algorithms and Models for the Web-Graph: Proceedings of the 5th International Workshop*, Berlin: Springer, 56–67.

Mlekuz D. (2012), 'Everything flows: Computational approaches to fluid landscapes', *CAA2012 Proceedings of the 40th Conference in Computer Applications and Quantitative Methods in Archaeology*, Southampton.

Mollot L. A. and R. E. Bilby (2008), 'The use of geographic information systems, remote sensing, and suitability modelling to identify conifer restoration sites with high biological potential for anadromous fish at the Cedar River Municipal Watershed in western Washington, USA', *Restoration Ecology* 16, 336–47.

Monterroso Checa A. (2017), 'Remote sensing and archaeology from Spanish LiDAR-PNOA: Identifying the amphitheatre of the Roman city of Torreparedones (Córdoba-Andalucía-Spain)', *Mediterranean Archaeology and Archaeometry* 17, 15–22.

Moreno Gallo I. (2006), *Vías romanas: Ingeniería y técnica constructive*, Madrid: Traianus.

Moreno-Navarro F. (2022), 'Simulación y análisis del sistema de transporte en el norte de la Carpetania romana: Una aproximación desde SIG y análisis de redes', *Zephyrus* 89, 191–211.

Moss G. (2020), *Hostile Ground: Terrain, Force and Control in the Roman Empire*, Chapel Hill: University of North Carolina, Doctoral Thesis.

Murray W. M. (1987), 'Do modern winds equal ancient winds?', *Mediterranean Historical Review* 2 (2), 139–67.

Murrieta Flores P., (2012), 'Understanding human movement through spatial technologies', *Trabajos de Prehistoria* 69 (1), 103–22.

Nolan K. and N. Cook (2012), 'A method for multiple cost-surface evaluation of a model of fort ancient interaction', in D. White and S. Surface-Evans (eds), *Least Cost Analysis of Social Landscapes: Archaeological Case studies*, University of Utah, 67–93.

Nutsford D., F. Reitsma, A. L. Pearson and S. Kingham (2015), 'Personalising the viewshed: Visibility analysis from the human perspective', *Applied Geography* 62, 1–7.

Olivella J. (1994), Teoría *del buque: Flotabilidad y estabilidad*, Barcelona: Ediciones UPC.

Olson L. (1943), 'Columella and the beginning of soil science', *Agricultural History Society* 17, 65–72.

Oltean I. A. and F. Fonte (2020), 'GIS analysis and spatial networking patterns in upland ancient warfare: The Roman conquest of Dacia', *Geosciences* 11 (1), 17.

Orengo H. A. and A. Livarda (2016), 'The seeds of commerce: A network analysis-based approach to the Romano-British transport system', *Journal of Archaeological Science* 66, 21–35.

Ormsby T., E. Napoleon, R. Burke, C. Groessl and L. Feaster (2004), *Getting to Know ArcGIS Desktop: Basics of ArcView, ArcEditor and ArcInfo*, Redlands, CA: ESRI Press.

Otavio Rodríguez E. (1984), *Extracción de trozas mediante bueyes y tractores agrícolas*, Roma: Estudio Fao Montes.

Parcero Oubiña C. (2013), 'Midiendo decisiones locacionales: Una aproximación a la evaluación de la defensibilidad efectiva de sitios arqueológicos fortificados', *Comechingonia: Revista de arqueología* 17, 57–2.

Parcero Oubiña C., A. Güimil Fariña, J. E. Fonte and J. M. Costa Garcia, (2019), 'Footprints and cartwheels on a pixel road: On the applicability of GIS for the modelling of ancient (Roman) routes BT', in P. Verhagen, J. Joyce, M. Groenhuijzen (eds), *Finding the Limits of the Limes*, Berlin: Springer, 291–311.

Pardo Abad C. (2008), *Turismo y patrimonio industrial: un análisis desde la perspectiva territorial*, Madrid: Síntesis.

Pasquini B. (2017), 'Preposition of a method to determine portages routes: the example of the connection between the Saône and the Moselle rivers (France) in ancient times', in R. O'Sullivan, C. Marini and J. Binnberg (eds), *Archaeological Approaches to Breaking Boundaries: Interaction, Integration and Division*, Oxford: British Archaeological Reports, 2–12.

Peeples M. A. (2019), 'Finding a place for networks in archaeology', *Journal of Archaeological Research* 27 (4), 1–49.

Perakis K. G. and A. K. Moysiadis (2011), 'Geospatial predictive modelling of the neolithic archaeological sites of Magnesia in Greece', *International Journal of Digital Earth* 4 (5), 421–33.

Petrasova A., B. Harmon, V. Petras, P. Trabrizian and H. Mitrasova (2018), *Tangible Modelling with Open-Source GIS*, Boston: Springer, 95–106.

Poirier N., R. Opitz, L. Nuninger and K. Ostir (2013), 'Lidar in Mediterranean agricultural landscapes: Reassessing land use in the Mauguio', *HAL, Archives-Ouvertes*, 1–17.

Polla S. and P. Verhagen (2014), *Computational Approaches to the Study of Movement in Archaeology: Theory, Practice and Interpretation of Factors and Effects of Long Term Landscape Formation and Transformation*, Vol. 23, Berlin: De Gruyter.

Pomey P. (2020), 'Naval architecture. The Hellenistic hull design: Origin and evolution', in E. Nantet (ed.), *Sailing From Polis to Empire: Ships in the Eastern Mediterranean during the Hellenistic Period*, Cambridge: Open book publisher, 27–54.

Premo L. S. (2004), 'Local spatial autocorrelation statistics quantify multi-scale patterns in distributional data: An example from the Maya lowlands', *Journal of Archaeological Science* 31 (7), 855–66.

Quirós E., P. Trapero Fernández, A. Antolín and V. Mayoral (2024) 'A comparative approach to GIS modelling of terrestrial mobility in archaeological sites: The Iron Age hillfort of Villasviejas del Tamuja as a study case', *Journal of Archaeological Science* 31, 102312.

Raepsaet G. (2002), *Attelages et techniques de transport dans le monde gréco-romain*, Bruxelles: Livre Timperman.

Rathbone D. (1991), *Economic Rationalism and Rural Society in Third-Century A.D. Egypt: The Heroninos Archive and the Appianus States*, Cambridge: Cambridge University Press.

Rawat N. S., T. Brughmans, V. Nautiyal and D. D. Chauniyal (2021), 'Networked medieval strongholds in Garhwal Himalaya, India', *Antiquity* 95 (381), 753–72.

Rees S. E. (1979), *Agricultural Implements in Prerromano and Roman Britain*, Oxford: British Archaeological Reports, British Series 69.

Remesal Rodríguez J. (1986), *La annona militaris y la exportación de aceite Bético a Germania*, Madrid: Universidad Complutense.

Richards Rissetto H. (2017), 'An interactive 3D GIS analysis of the role of visibility in ancient Maya landscapes: a case study form Copan, Honduras', *Digital Scholarship in the Humanities* 32 (2), 195–212.

Rivas I. B. (2019), *GIS and Marble Transportation in the Roman Imperial Period: Modelling a Journey from MT. Pyrgari (Euboea, Greece) to Rome*, Amsterdam: Vrije Universiteit Amsterdam.

Rivers R., C. Knappett and T. Evans (2013), 'What makes a site important? Centrality, gateways and gravity', in C. Knappett (ed.), *Network Analysis in Archaeology: New Approaches to Regional Interaction*, Oxford: Oxford University Press, 125–50.

Rosenstein S. (2004), *Rome at War, Farms, Families an Death in the Middle Republic*, University of North Carolina Press.

Ruzickova K., J. Ruzicka and J. Bitta (2021), 'A new GIS-compatible methodology for visibility analysis in digital surface models of earth sites', *Geoscience Frontiers* 12, 4.

Saaty T. L. (1994), *Fundamentals of Decision Making and Priority Theory with the Analytic Hierarchy Process*, Pittsburgh: RWS Publications.

Sáez Fernández P. (1978), *Agricultura romana de la Bética*, Sevilla: Departamento de Historia Antigua, Universidad de Sevilla.

Sáez Fernández, P. (1988), 'Sobre algunos tipos de tierras vitícolas de la Bética', *Actas I congreso peninsular de Historia Antigua*, Coimbra, 521–31.

Safadi C. (2016), 'Wind and wave modelling for the evaluation of the maritime accessibility and protection afforded by ancient harbours', *Journal of Archaeological Science: Reports* 5, 348–60.

Sánchez Aparicio L., M. G. Masciotta, J. García Alvarez, L. F. Ramos, D. V. Oliveira, J. A. Martín Jiménez and P. Monteiro (2020), 'Web-GIS approach to preventive conservation of heritage buildings', *Automation in Construction* 118, 103304.

Sánchez-Simón M. (2021), '*Villae* and Farms: Early Imperial Rural Settlement in the Adaja-Eresma Basin (Central Roman Spain)', in J. Bermejo Tirado and I. Grau Mira (eds), *The Archaeology of Peasantry in Roman Spain*, Leck: De Gruyter, 201–28.

Sarmady S., F. Haron and A. Talib (2014), 'Simulation of pedestrian movements on slope using fine grid cellular automata', *IAES International Journal of Artificial Intelligence* 11 (4), 1197–212.

Sarris A. (2015), *Best Practices of GeoInformatic Technologies for the Mapping of Archaeolandscapes*, Oxford: Archaeopress Archaeology.

Sarris I. M., D. Chalkiadakis and A. Sarris (2022), 'Investigating an automatic segmentation and classification of archaeological magnetic features through machine learning techniques', in *Proceedings of the 27th International Conference on Cultural Heritage and New Technologies*, Heidelberg.

Scheidel W. (2008), 'Roman population size: the logic of the debate', in L. De Light and S. Northwood (eds), *People, Land, and Politics: Demographic Developments and the Transformation of Roman Italy 300 BC – AD 14*, Leiden: Brill, 17–70.

Scheidel W. (2014), 'The shape of the Roman world: modelling imperial connectivity', *Journal of Roman Archaeology* 27, 7–32.

Scheidel W. and E. Meeks (2012), 'ORBIS: The Stanford geospatial network model of the Roman World', Available online: http://orbis.stanford.edu (accesed 20 December 2022).

Scianna A. and B. Villa (2011), 'GIS applications in Archaeology', *Archeologia e calcolatori* 22, 337–63.

Shi F., C. Sun, A. Guion, Q. Yin, S. Zhao, T. Liu and Z. Guo (2022), 'Roman Warm Period and Late Antique Little Ice Age in an earth system model large ensemble', *Journal of Geophysical Research: Atmospheres* 127, 10.1029/2021JD035832.

Sibson R. (1981), 'A brief description of natural neighbour interpolation', in V. Barnett (ed.), *Interpreting Multivariate Data*, New York: John Wiley & Sons, 21–36.

Sillières P. (1990), 'La búsqueda de las calzadas romanas: desde la foto-interpretaación hasta el sondeo', *Simposio sobre la red viaria en la Hispania romana*, Zaragoza: Diputación provincial de Zaragoza.

Smith C. and E. E. Cochrane (2011), 'How is visibility important for defence? A GIS analysis of sites in the western Fijian Islands', *Archaeology in Oceania* 46, 76–84.

Steinberg J. (1996), 'Ploughzone sampling in Denmark: isolating and interpreting site signatures from disturbed contexts', *Antiquity* 70, 368–92.

Stek T. (2014), 'The city-state model and the Roman Republican colonization: sacred landscapes as proxy for colonial socio-political organisation', in T. Stek and J. Pelgrom (eds), *Roman Republican Colonization: New Perspectives from Archaeology and Ancient History*, Rome: Papers of the Royal Netherlands Institute in Rome. Palombi Editori, 87–105.

Stöger H. (2011), *Rethinking Ostia: A Spatial Enquiry into the Urban Society of Rome's Imperial Port-Town*, Archaeological Studies Leiden University 24, Leiden: Leiden University Press.

Store R. and J. Jokimaki (2003), 'A GIS-based multi-scale approach to habitat suitability modelling', *Ecological Modelling* 169, 1–15.

Stott D., S. M. Kristiansen and S. M. Sindbæk (2019), 'Searching for Viking Age fortresses with automatic landscape classification and feature detection', *Remote Sensing* 11 (16), 1881.

Stubert L., A. Martín, I. Oliveras, M. Märker, H. Schernthanner and S. Vogel (2020), 'Viticulture in the Laetanian region (Spain during the Roman Period: Predictive modelling and geomatic analysis', *Geociences* 10 (6), 206.

Surface-Evans S. (2012), 'Cost catchments: A least cost application for modeling hunter-gatherer land use', in D. White and S. Surface-Evans (eds), *Least Cost Analysis of Social Landscapes: Archaeological Case Studies*, University of Utah, 128–51.

Tartaron T. F. (2014), *Maritime Networks in the Mycenaean World*, Cambridge: Cambridge University Press.

Tchernia A. (1986), *Le vin de l'italie romaine*, Rome: Ecole de francaise de Rome.

Tilley C. (2010), *Interpreting Landscapes: Geologies, Topographies, Identities; Explorations in Landscape Phenomenology 3*, Walnut Creek: Left Coast Press.

Tobler W. (1993), 'Three presentations on geographical analysis and modeling: non-isotrophic geographic modeling; speculations on the geometry of geography; and global spatial analysis', *Technical Report 93-1*, Santa Barbara: National Center for Geographic Information and Analysis, University of California. Retrieved from https://escholarship.org/uc/item/05r820mz.

Trapero Fernández P. (2016), 'Roman viticulture analysis based on latin agronomist and geographic information systems in lower Guadalquivir', *Virtual Archaeology Review* 7 (14), 53–60.

Trapero Fernández P. (2019), 'Conectividad en el estuario del Guadalquivir entre Turris Caepionis y Nabrissa Veneria: Aprovechamientos económicos, comunicaciones, embarcaderos y zonas navegables', in L. Lagóstena Barrios (ed.) *Prácticas sostenibles y aprovechamientos históricos*, Barcelona: Barcelona University, Instrumenta serie, 179–95.

Trapero Fernández P. (2021a), *La viticultura romana en el estuario del Guadalquivir: Las prácticas de cultivo, producción, distribución y modelado SIG en la colonia Hasta Regia*, Oxford: British Archaeological Reports, International Series. S3022.

Trapero Fernández P. (2021b), 'Mobility in Ancient Times: combining land and water costs', *Digital Applications in Archaeology and Cultural Heritage*, 22 (6) e00192.

Trapero Fernández P. (2021c), 'Condicionantes para el cultivo de la vid en época romana a través de Columela y los Sig', *Revista de estudios Andaluces* 41, 25–46.

Trapero Fernández P. (2022), 'Estudio de movilidad en el transporte romano de los mármoles de Estremoz', *Mármore 2000 anos de História: Projecto internacional de Historia da Indústria dos Mármores*, Vol. III, 43–65.

Trapero Fernández P. (2023), 'Modelo predictivo de aprovechamientos vitivinícolas: La colonia romana de Hasta Regia, Hispania', *Archaeologia e Calcolatori* 34 (1), 311–20.

Trapero Fernández P. and A. Carneiro (2023), 'GIS Modelling of surveyor conditions: Method for identifying fundi in Alto Alentejo (Portugal)', *Journal of Archeological Science Reports* 51, 104212.

Trapero Fernández P. and A. Carneiro (eds) (2023), *GIS Applications in Roman Landscape and Territory: Methodologies and Models in Hispania*, Oxford: British Archaeological Reports, International Series.

Trapero Fernández P. and E. Aragón Núñez (2022), 'Modelling Cabotage: Coastal navigation in the western Mediterranean Sea during the Early Iron Age', *Journal of Archaeological Science Reports* 41, 103270.

Trapero Fernández P. and J. A. Ruiz Gil (2023), 'Estudio SIG de los lugares defensivos de forma diacrónica en la comarca de La Janda (provincia de Cádiz)', *Arqueología y territorio medieval* 30, 261–85.

Trapero Fernández P. and L. Lagóstena Barrios (2019), 'La localización de los viñedos de Marco Columela: literatura agronómica y análisis GIS en el interfluvio Guadalquivir-Guadalete', in Rodríguez, J. R., et al. (eds), *Paisajes productivos y redes comerciales en el imperio romano*, Barcelona: Barcelona University, Instrumenta 65, 89–114.

Trapero Fernández P., A. Carneiro and N. Moreira (2023a), 'Transport and distribution of heavy loads in ancient times: Estremoz Marbles in the Roman province of Lusitania', *Journal of Archaeological Science Reports* 49, 103962.

Trapero Fernández P., J. Catalán González and E. Ruiz Pilares (2023b), 'Modelado histórico del Río Guadalete con SIG: Restitución diacrónica del cauce y los condicionantes de navegabilidad', *Vínculos de Historia* 12, 261–84.

Tukey J. (1977), *Exploratory Data Analysis*, Reading: Addison-Wesley.

Turchetto J. and G. Salemi (2017), 'Hide and Seek: Roads, lookouts and directional visibility cones in Central Anatolia', *Open Archaeology* 3, 69–82.

Van Leusen M. (1999), 'Viewshed and cost surface analysis using GIS (cartographic modelling in a cell-based GIS II, New Techniques for Old Times', *CAA 98 Computer Applications and Quantitative Methods in Archaeology: Proceedings of the 26th Conference*, Oxford: British Archaeological Reports, 215–23.

Van Leusen M. (2002), *Pattern to Process: Methodological Investigations into the Formation and Interpretation of Spatial Patterns in Archaeological Landscapes*, Groninga: Rijksuniversiteit Groningen.

Vaughn S. and T. Crawford (2009), 'A predictive model of archaeological potential: An example from northwestern Belize', *Applied Geography* 29, 542–55.

Verboven K. and C. Laes (2016), *Work, Labour, and Professions in the Roman World*, Leiden Boston: Brill.

Verbrugghe G., W. De Clercq and V. Van Eetvelde (2017), 'Routes across the Civitas Menapiorum: Using least cost paths and GIS to locate the Roman roads of Sandy Flanders', *Journal of Historical Geography* 57, 76–88.

Verhagen P. (2005), 'Quantifying the qualified: The use of multicriteria methods and Bayesian statistics for the development of archaeological predictive models', in M. Mehrer and K. L. Wescott (eds), *GIS and Archaeological Site Location Modeling*, Boca Raton: CRC Press, 191–216.

Verhagen P. (2007), 'Quantifying the qualified: the use of multicriteria methods and Bayesian statistics for the development of archaeological predictive models', in Verhagen, P., *Case Studies in Archaeological Predictive Modelling*, Amsterdam University Press, 71–91.

Verhagen P. (2018), 'Spatial analysis in archaeology: Moving into new territories BT', in C. Siart, M. Forbriger and O. Bubenzer (eds), *Digital Geoarchaeology: New Techniques for Interdisciplinary Human-Environmental Research*, Berlin: Springer, 11–25.

Verhagen P., J. Joyce and M. Groenhuijzen (2014), 'Modelling the dynamics of demography in the Dutch Roman limes zone', *Landscape Archeology Conference 2014 Proceedings*, 1–13.

Verhagen P., L. Nuninger and M. R. Groenhuijzen (2019), 'Modelling of pathways and movement networks in archaeology: An overview of current approaches', in P. Verhagen, J. Joyce and R. Mark (eds), *Finding the Limits of the Limes: Computational Social Sciences*, New York: Springer, 217–49.

Verhagen P., T. Brughmans, L. Nuninger and F. Bertoncello (2013), 'The long and winding road: Combining least cost paths and network analysis techniques for settlement location analysis and predictive modelling', in I. Romanowska, P. Murrieta Flores, C. Papadopoulos and A. Chrysanthi (coords), *Proceedings of the*

40th Conference in Computer Applications and Quantitative Methods in Archaeology, Southampton: CAA International, 357–66.

Viitanen E. M. (2010), *Locus Bonus: the Relationship of the Roman Villa to Its Environment in the Vicinity of Rome*, Helsinki: Helsinki University Press.

Villalba Babiloni T. E. (2018), 'Embarcaciones y navegantes fluviales al servicio de los antiguos puertos de Roma', in *VI Congreso virtual sobre Historia de las Vías de Comunicación*, Jaén: Asociación Orden de la Caminería, 475–88.

Vivoni E. R. and K. T. Richards (2005), 'Integrated use of GIS-based field sampling and modeling for hydrologic and water quality studies', *Journal of Hydroinformatics* 7, 235–50.

Vogel S., M. Maerker, D. Esposito and F. Seiler (2016), 'The ancient rural settlement structure in the *hinterland* of Pompeii inferred from spatial analysis and predictive modeling of *villae rusticae*', *Geoarchaeology* 31, 121–39.

Voorrips A. (1987), 'Formal and statistical models in archaeology', *Quantitative Research in Archaeology: Progress and Prospects* 3, 61–72.

Wachtel I., R. Zidon, S. Garti and G. Shelach-Lavi (2018), 'Predictive modeling for archaeological site locations: Comparing logistic regression and maximal entropy in north Israel and north-east China', *Journal of Archaeological Science* 92, 28–36.

Wang X. H. and Z. Y. Yin (1997), 'Using GIS to assess the relationship between land use and water quality at a watershed level', *Environment International* 23, 103–14.

Watkins R. (2015), 'Terrain metrics and landscape characterization from bathymetric data: SAGA GIS methods and command sequences', *Report Prepared for the Ecospatial Information Team, Coral Reef Ecosystem Division, Pacific Islands Fisheries Science Center: Honolulu, HI, under NOAA Contract.*

Weber M. (2008), *Historia agraria romana*, Madrid: Akal editorial.

Weiss A. (2000), *Topographic Position and Landforms Analysis*. http://www.jennessent.com/downloads/tpi-poster-tnc_18x22.pdf (accessed 10 of Septembre 2019).

Werther L. and L. Kröger (2017), 'Medieval inland navigation and the shifting fluvial landscape between Rhine and Danube (Germany)', *Post-Classical Archaeologies* 7, 65–96.

Wheatley D. (1995), 'Cumulative viewshed analysis: a GIS-based method for investigating intervisibility, and its archaeological application', in G. Lock and Z. Stančič (eds), *Archaeology and GIS: A European Perspective*, London: Routledge, 171–86.

Wheatley D. and M. Gillings (2002), *Spatial Technologies in Archaeology*, London: Taylor & Francis.

White D. A. (2015), 'The basics of least cost analysis for archaeological applications', *Advances in Archaeological Practice* 3 (4), 407–14.

Wilson J. P. and J. C. Gallant (2000), *Terrain Analysis: Principles and Applications*, London: John Wiley & Sons.

Witcher R. E. (2008), 'Resurveying Mediterranean rural landscapes: GIS and legacy survey data', in *Internet Archaeology* 24, 10.11141/ia.24.2.

Yin P. (2020), 'Kernels and density estimation', in J. P. Wilson (ed.), *The Geographic Information Science & Technology Body of Knowledge* (1st Quarter 2020 Edition).

Zamora Merchán M. (2006), 'Visibilidad y SIG en Arqueología: mucho más que ceros y unos', in I. Grau Mira (ed.), *La Aplicación de los SIG en la Arqueología del Paisaje*, Alicante: Universidad de Alicante, 41–54.

Zanni S. and A. De Rosa (2018), 'Remote sensing analyses on Sentinel-2 images: Looking for Roman roads in Srem region (Serbia)', *Geosciences* 9 (1), 25.

Zevenbergen L. W. and C. Thorne (1987), 'Quantitative analysis of land surface topography', *Earth Surface Processes and Landforms* 12, 47–56.

Zheng P. Q. and B. W. Baetz (1999), 'GIS-based analysis of development options from a hydrology perspective', *Journal of Urban Planning and Development-Asce* 125, 164–80.

Zhou Q. (2017), 'Digital elevation model and digital surface model', in D. Richardson, N. Castree, M. F. Goodchild, A. Kobayashi, W. Liu and R. A. Marston (eds), *The International Encyclopedia of Geography*, Chichester: John Wiley & Sons, Ltd.

Zhu J., Y. Nie, H. Gao, F. Liu and L. Yu (2017), 'GIS-based visibility network and defensibility model to reconstruct defensive system of the Han dynasty in central Xinjiang, China', *ISPRS International Journal of Geo-Information* 6, 247.

Zorn J. (1994), 'Estimating the population size of ancient settlements: Methods, problems, solutions, and a case study', *BASOR* 295, 31–48.

Index

www.ingramcontent.com/pod-product-compliance
Ingram Content Group UK Ltd.
Pitfield, Milton Keynes, MK11 3LW, UK
UKHW020258051026
15470UKWH00020B/1097

9 781350 433700